Monitoring Movements in Development Aid

Infrastructures Series
edited by Geoffrey Bowker and Paul N. Edwards

Paul N. Edwards
A Vast Machine: Computer Models, Climate Data, and the Politics of Global Warming

Lawrence M. Busch
Standards: Recipes for Reality

Lisa Gitelman, ed.
"Raw Data" Is an Oxymoron

Finn Brunton
Spam: A Shadow History of the Internet

Nil Disco and Eda Kranakis, eds.
Cosmopolitan Commons: Sharing Resources and Risks across Borders

Casper Bruun Jensen and Brit Ross Winthereik
Monitoring Movements in Development Aid: Recursive Partnerships and Infrastructures

Monitoring Movements in Development Aid

Recursive Partnerships and Infrastructures

Casper Bruun Jensen and Brit Ross Winthereik

The MIT Press
Cambridge, Massachusetts
London, England

MIT Press books may be purchased at special quantity discounts for business or sales promotional use. For information, please email special_sales@mitpress.mit.edu or write to Special Sales Department, The MIT Press, 55 Hayward Street, Cambridge, MA 02142.

Set in Stone Sans and Stone Serif by the MIT Press. Printed and bound in the United States of America.

Library of Congress Cataloging-in-Publication Data

Jensen, Casper Bruun.
Monitoring movements in development aid : recursive partnerships and infrastructures / Casper Bruun Jensen and Brit Ross Winthereik.
pages cm. — (Infrastructures series)
Includes bibliographical references and index.
ISBN 978-0-262-01965-1 (hardcover : alk. paper)
1. Economic assistance—Information technology. 2. Economic development projects—Evaluation. 3. Information technology—Economic aspects. 4. Infrastructure (Economics) 5. Economic development. I. Winthereik, Brit Ross, 1973– II. Title.
HC60.J43 2013
338.91—dc23
2013005453

10 9 8 7 6 5 4 3 2 1

for Moeko and Thomas

Contents

Preface

As is always the case, this book is the result of many kinds of conversations with many different people.

For their critical and constructive comments at workshops at Osaka University, at the University of Copenhagen, and at the IT University of Copenhagen, we thank our colleagues in the Environmental Infrastructures research project: Anders Blok, Shuhei Kimura, Atsuro Morita, Osamu Nakagawa, Moe Nakazora, and Moeko Saito-Jensen.

Bill Maurer, Stefan Helmreich, Mei Zhan, and the participants in the 2011 Sideways Workshop offered valuable feedback on recursions (chapter 2), as did participants in the 2009 conference of the Danish Association for Science and Technology Studies and participants in the 2009 European Science Foundation workshop on Mapping Interfaces: The Future of Knowledge. For giving Casper the opportunity to join the latter event we thank Gísli Pálsson, and we thank Marilyn Strathern for the recommendation!

"Inventive Frontiers" (chapter 3 here) was presented at the 4S/EASST Conference in Washington in 2009, at the workshop on Intercultural Communication at the IT University in 2010, in the lecture series on Material Culture and Social Anthropology at University College London in 2010, at the Department of Informatics of Oslo University in 2010 as part of the Global Infrastructures research project, and in an advanced anthropology graduate seminar at Tokyo University in 2011. Warm thanks to Margunn Aanestad, Pernille Bjørn, Martin Holbraad, Göran Sundqvist, and Tadashi Yanai for inviting us to those exciting events and to Hirotsugu Terado and Yuichi Sekiya for being discussants in Tokyo. Special thanks to Rob Hagendijk and Michael Lynch for helping us to present at the 4S conference in Washington, where our abstract initially suffered the indignity of being rejected as "irrelevant to STS."

"Development Loop" (chapter 4) benefited from careful reading by Christopher Gad and Bjarki Valtysson.

"Weedy Infrastructure" (chapter 5) was presented in 2011 at the Environmental Infrastructures workshop at Osaka University, at the Netherlands Graduate School of Science, Technology, and Modern Culture (WTMC), at the Center for Network Culture at the IT University, and at the Data–Difference–Diversity workshop at the Max Planck Institute for the History of Science. We are grateful to all who helped us with this chapter. We offer special thanks to Susanne Bauer, Geoff Bowker, Willem Halffman, Christine Hanke, Sabine Höhler, Irina Shklovski, T. L. Taylor, and Teun Zuiderent-Jerak.

"Wormholes" (chapter 6) was presented to the STS Network Japan at Tokyo University in 2011. Thanks so much to Arisa Ema for organizing this event, to Ryuma Shineha, Mikami Kouichi, and Kenji Ito for feedback, and to Sho Morishita and Wakana Suzuki for showing up.

The research that led to this book did not benefit from dedicated external funding. Thus, we are particularly appreciative of the early investment in our project provided by Provost Jørgen Staunstrup at the IT University of Copenhagen and of later support to field studies provided by the ITU. We are grateful for the assistance provided by the ITU research administration (especially Nhi Quyen Le, who helped us organize the first visit to Vietnam). Casper would also like to express his gratitude to the DKUNI–Japanese Society for the Promotion of Science exchange program for grant 2010-3411-05, which funded a three-month research visit to Osaka University's Department of Anthropology. Very special thanks to Atsuro Morita for being such a wonderful host, friend, and intellectual sparring partner.

Helen Verran deserves particularly heartfelt thanks for welcoming Brit at the University of Melbourne's School of Philosophy, Anthropology, and Social Inquiry during a two-month research visit in 2009.

We are extremely grateful to Anne Beaulieu and Laura Watts, both of whom read the entire manuscript on short notice. Thanks very much to Paul Edwards and Geoff Bowker for encouraging us to submit the manuscript to the Infrastructures series, to Margy Avery for being a wonderful editor, to Katie Persons for her friendly and professional help, and to Paul Bethge for his dedication to making the book readable.

Aside from these specific debts, we thank the following people for sharpening our thinking about monitoring movements and infrastructure, about science and technology studies, ethnography, anthropology and their connections and disconnections, and about the conceptual and the empirical: Nina Boulus, Karen Boll, Nathalia Brichet, Steve Brown, Eduardo Viveiros de Castro, Christopher Gad, Birgitte Gorm Hansen, Torben Elgaard Jensen, Susanne Branner Jespersen, Naoki Kasuga, Henriette Langstrup, Annegrete

Juul Nielsen, Martin Holbraad, Morten Nielsen, Morten Axel Pedersen, Dixi Strand, Lucy Suchman, Lotte Thomsen, Soumhya Venkatesan, Signe Vikkelsø, and Antonia Walford. Warm thanks to all former and present colleagues in the Technologies in Practice faculty group, and to former and present students—particularly Tina Skjærbæk Søborg and other students in Brit's Situational Methods course who helped with some of the early interviewing and transcribing.

We would like to offer our sincere thanks to all who have been willing to share their views and experiences with creating accountability, transparency, and partnership through and against infrastructure, including employees of the Danish Development Agency (Danida), employees of the Danish Embassy in Hanoi, employees of the Danish National Audit Office, and members of the AidData project. We would also like to thank the independent international development consultants who took time out of their busy schedules to let themselves be interviewed, and especially the interviewees involved with Publish What You Fund and the International Aid Transparency Initiative. We are indebted to the managers and employees at NatureAid who patiently let us participate in meetings and who talked to us while working on many other urgent matters. Special thanks to Lasse Juul-Olsen for helping us to understand the practices of monitoring in environmental work and for encouraging us to do fieldwork with NatureAid in the first place. We have anonymized all people and most organizations referred to in the book, with a few exceptions where we have received explicit permission or where the material to which we refer is directly available in the public domain.

Chapter 2 was published in significantly different form as "Recursive partnerships in global development aid" in *Differentiating Development: Beyond an Anthropology of Critique*, edited by Soumhya Venkatesan and Thomas Yarrow (Berghahn, 2012).

Introduction

This is a book about emerging aid information infrastructures constructed to enhance modes of creating accountability and of ensuring effectiveness in aid development. It is about technologies developed to monitor organizations, practices, or projects involved in aid and to make information accessible and transparent to a broad range of global users. It is also about aid partnership as an ideal and as a collaborative form. It is at once about these empirical phenomena and about the analytical and methodological challenges they pose for science and technology studies (STS) and for social and cultural anthropology.

We describe and discuss a number of initiatives, places, organizations, and people involved in the monitoring of development aid. Across these practices, people are concerned with how to make better use of information technology in order to improve accountability in aid practices. The sites and actors we study differ considerably, and they are not all connected. We designate this loosely coupled global assemblage (Ong and Collier 2005), which some actors would like to see much more integrated, *monitoring movement*, a term that provides us with a way of keeping the empirical and the conceptual at the same level.

In one sense, information infrastructures, monitoring, transparency, and partnerships are parts of the same set of activities and ways of thinking. Without the strong contemporary focus on partnership, aid information infrastructures would look quite different. But then, without information infrastructures that give form to ideas about transparency and accountability the notion and ideal of partnership, too, would differ. However, this is only one side of the story. For in another sense partnership provides a collaborative form encompassing all these different activities. This is important for everything that follows, because partnerships create relations of recursive implication. In a context in which almost everything can be conceived as a partnership, relational loops between widely differing actors are rampant. Such loops define what counts as knowledge within partnerships, what counts as transparency, and even what counts as partnership itself. In this sense, partnership is a self-contextualizing form. This is an empirical point, but it is also of conceptual and methodological consequence. Engaging the monitoring movement, we suggest, entails conceptual and methodological invention (see Jensen 2012) and a rethinking of dominant understandings of what is involved in, and entailed by, the knowledge-making practices of STS and social anthropology.

We begin our exploration of aid information infrastructures by pointing attention to a document that aroused our curiosity: the Paris Declaration on Aid Effectiveness. The declaration (available at http://www.oecd.org), signed by 111 countries in 2005, "lays out a practical action-orientated roadmap to improve the quality of aid and its impact on development." It was organized around five principles: "ownership, alignment, harmonization, managing for results, and mutual accountability." At its heart was a wish to commit donors and receiving countries to making their strategies, their plans, and their actions transparent to one another. Among other things, the declaration committed donors to working toward standardized monitoring and evaluation systems so that partnering governments would no longer be subject to reporting styles and formats defined by each individual donor. Moreover, the signing donors agreed to work toward aligning their funding strategies with the strategies and budgets of the receiving countries. Governments in the countries receiving aid, meanwhile, agreed to make budgets and planned decisions more transparent to donors and more open to public scrutiny in their domestic environments.

We were fascinated by this image of the future of aid (see Anders 2005) and by questions of infrastructure that arise once one begins to think about new mechanisms of accountability and transparency. Our first idea centered on the notion of a *chain of accountability*. How, we wondered, can agencies located in Denmark, in Japan, or in the United Kingdom know that money spent on projects in Vietnam or in India has been spent well? (See also Rottenburg 2009.) How can they know that their activities and projects in other parts of the world run smoothly and achieve their objectives? Central to this knowledge is *aid information* about project progress and financial flows. Such information is found in spreadsheets, indicators, charts, databases, and registries, but also in encounters and meetings that never make it into formal infrastructures.

Thus, we became interested in exploring what would be required for aid information to remain intelligible and relevant as it moved from development projects to donor monitoring and evaluation units. We imagined focusing on how reporting practices create paths of visibility between project sites, development agencies, and funders. Our initial ambition was to trace information undergoing a series of transformations as it moved through the infrastructures used for development aid (henceforth referred to as *aid infrastructures*).

To our surprise, many development practitioners easily recognized "our" concept of "the chain of accountability." At the outset, we thought this was an advantage, because it enabled us to establish a degree of mutual understanding with employees in development organizations and non-governmental organizations such as the Danish Development Agency (Danida) and NatureAid. But the fact that some of our informants literalized the metaphor of the chain also meant that our ambition to engage it analytically created partnership hindrances. Thus, some of our potential informants saw our ethnographic and conceptual endeavor as not sufficiently useful to be worth their time.

In the event, the book isn't concerned with tracing information through chains of accountability. Instead, it is concerned with articulating the complex infrastructural, social, and cultural processes that commitments to accountability and transparency have set in motion, and with analyzing some of the varied practical and conceptual stakes of those processes. Exploring monitoring initiatives and technologies in a very diverse set of cases, we show infrastructure to be central to these practices and the ambitions they express. Crucially, making aid infrastructures is not simply a matter of designing and implementing *technology*. Making aid infrastructures entails forging new platforms for action that are simultaneously imaginative and practical, simultaneously conceptual and technical. Studying monitoring movements entails studying this work as well as attending to what falls through the cracks of infrastructure.

Intellectually, the book is in conversation with several bodies of research. First and foremost, it is located within STS, a field in which infrastructures have become increasingly important sites of investigation (Wouters et al. 2008) as a consequence of the more general interest in analyzing the social dimensions of technology. Infrastructure has become a prominent matter of concern because large-scale scientific collaboration, global business and enterprise, distributed organizational practices of technology development, international policy making, and worldwide communication networks—that is, multiple forms of infrastructure—are important to those interested in understanding how science and technology operate today.

But while this book is about globally distributed efforts to enhance monitoring, auditing, and accountability practices in aid development, it is also about new social practices and cultural imaginations to which such projects give rise in settings all over the world. This is why, in our empirical setting, infrastructure studies connect with many concerns that are central to social and cultural anthropology, and particularly to the anthropology of development. Indeed, the anthropology of development is itself undergoing change, in part as a consequence of encounters with STS modes of conceptualization and analysis (Yarrow and Venkatesan 2012). The analytical movement we outline, then, focuses on how STS research has engaged with infrastructure studies in adjacent fields, and on how development aid practices have begun to be conceptualized by an anthropological literature that is itself increasingly conversant with STS. We add to this complex field of discussions a mode of engagement that we specify as practical ontology.

Chapter 1 aims to specify an analytical platform from which research on information infrastructures, science and technology studies, and the anthropology of development can be seen through each other. Engaging with a large body of literature from these fields, we aim to invent a space for translation and recalibration of research questions and agendas. Our deliberate use of the term *invention* indicates that our aspiration is not to combine insights from these fields in order to create a more representationally adequate (and holistic) perspective on aid infrastructures and monitoring movements. Rather, following Bruno Latour and Peter Weibel (2005), we would characterize our hope as one of "making a public."

By juxtaposing and interlinking ideas and approaches from the aforementioned fields of inquiry, we address imagined readers who might find inspiration in seeing discussions from their own fields diffracted through discussions taking place in adjacent fields. This aim might be qualified as a form of lateral movement (Maurer 2005; see also Jensen and Rödje 2010) in which concepts and ideas are made to flow across disciplinary boundaries in multiple directions and establish new points of intellectual contact. In particular, we specify our own mode of engagement with monitoring movements in terms of *practical ontology*, an approach that draws simultaneously on non-humanist STS research and multi-naturalist anthropology. For readers who are interested in conceptual issues at the intersection of STS and social anthropology, chapter 1 is the obvious starting point.

Readers more eager to engage with ethnographies of aid infrastructures might begin with chapter 2. The inventive aim of that chapter is to consider the recursive analytical and methodological implications of studying aid partnerships while also operating within them. In one sense, therefore, the chapter is broadly methodological. Yet this phrase is imprecise, since the chapter deals not only with issues of method but also with the recursive relations between empirical aid infrastructures and partnerships, our methods for engaging these sites, and our modes of conceptualization. The chapter takes the reader on a tour of some of our field sites. It describes the tense negotiations, carried out by Southeast Asian governments in preparation for the Accra High Level Forum on Aid Effectiveness, over the very meaning of partnership. Further, the chapter recounts episodes that had us partner with "research subjects" in ways that redefined our relations, our methods, our questions, and our forms of knowledge making. In these different cases, recursion entails that categories such as "the conceptual," "the empirical," and "the methodological" become "interactive kinds." (See Hacking 1986; Smith 1997.)

Chapter 3 offers empirical descriptions and analytical characterizations of several partially related emerging aid infrastructures for transparency and accountability. Not only are new digital platforms imagined in these settings; in addition, new forms of users and publics are at stake. Our empirical entry point is a conference on aid transparency and corruption—a conference that also saw the launch of the comprehensive AidData database. We use this conference diagnostically to characterize a set of real and imagined users and publics inhabiting what we refer to as the inventive frontiers of aid infrastructures. The chapter attunes us to the plurality of aid information infrastructure designs and their organizations, agendas, and imagined publics. We argue that both concrete and abstract users are crucial to the making of aid infrastructures.

Continuing the discussion from chapter 3, chapter 4 explores current efforts to build Web platforms and applications that seek to make existing datasets and information archives dynamic and available to the public. Delving into the Web-mediated techno-politics of the self-designated "transparency movement," we discuss an app competition that was held with the purpose of "activating" inert aid information held in the World Bank's data catalog. Such activation is currently seen to hold great promise for development policy and research, but what specific potential it holds remains in question. Our discussion of this issue extends into a discussion of how apps and platforms for aid transparency challenge conventional social scientific ways of conceptualizing the relation between digital infrastructures and political action.

In chapter 5, the setting changes from the inventive frontiers of aid technology to a setting within an already existing infrastructure. Thus, the chapter takes us to the monitoring and auditing activities of the Danish branch of NatureAid, an international environmental non-governmental organization. We follow the efforts of this NGO's conservation team to generate knowledge about a development project on ecotourism in a national park in Southern Vietnam. Describing events from a review mission carried out jointly by consultants from the Danish and Vietnamese offices of NatureAid, the chapter zooms in on the dilemmas of making knowledge in collaborative forms in which the issues of what partnership and knowledge mean are at stake. Processes of making partnership and concrete moments of partnership, we show, create weeds that grow in the interstices of formal monitoring infrastructure. Such weeds, we suggest, cannot and should not be held at bay. Paradoxically, they are both generative and destructive for infrastructures of aid accountability.

Chapter 6 explores performance audits conducted by the Danish National Audit Office. Among many other things, such performance audits are used to evaluate whether money from Danish taxpayers is spent appropriately by development aid projects. Surprisingly, we find that performance audits are also increasingly caught up in partnership relations. The National Audit Office is known for and proud of its procedural objectivity. At the same time, collaboration with audited organizations and mutual learning are both stated ambitions and practical requirements for performance auditing. The chapter scrutinizes the tensions inherent in the double obligation that auditing now is simultaneously about external control and inducing organizational learning. It shows how these simultaneous aspirations are rendered operational through a looping infrastructure that swirls in and out of audited organizations and the Danish public sphere. These audit loops, we argue, create a kind of wormhole (Haraway 1997) through which performance audit is morphing into new epistemological, social, and cultural shapes.

In the concluding chapter, we draw on the analyses presented in the preceding chapters in order to characterize the still emerging monitoring movement of contemporary development aid. This movement, we suggest, can be specified both in terms of certain shared themes and contents and in terms of particular forms and aesthetics. The monitoring movement comprises multiple areas of agency, parceled out and moving between multiple organizations and infrastructures, both imagined and concrete. Central to the monitoring movement is a collapse between means and ends, manifested in the ways in which aid information infrastructures are seen as at once technological platforms for making knowledge about aid *and* forms of aid and empowerment in their own right. However, while transparency advocates envision a future of real-time aid infrastructures for everyone, our analyses pay attention to the inherent patchiness and recursivity of such infrastructures. Contrary to what is often assumed, in aid development, as elsewhere, this patchiness is not so much a problem to be solved as it is a *fact* about infrastructure. We further argue that an important consequence of the increasing spread of aid infrastructures as a way of sharing information globally is the transformation of aid organizations and their employers into "modulating entities" created by the partial datasets to which they have access. Eventually the chapter returns to the question of what kinds of contributions practical ontology might make to STS and to social and cultural anthropology.

1 Infrastructures and Development Aid: Fields, Fractals, and Frictions

Infrastructures as Controversial Objects

Infrastructures are crucial to the operations of modern society and its organizations. They are also tricky analytical objects. The historian of technology Paul Edwards reminds us that *infrastructure* was originally a military term designating "fixed facilities" (2003, 186). Infrastructure was seen as the basis on which, or the means by which, a society or an organization operated; it included road networks, waterways, power grids, and even schools and prisons (ibid., 187, quoting the *American Heritage Dictionary*).

Today, however, infrastructures are widely perceived to have become less fixed and more fluid. This has to do with the fact that infrastructures are increasingly used for the production, distribution, and sharing of information. For example, aid practitioners have begun experimenting with crowdsourcing technologies to hold donors accountable for the quality of their programs and projects. Information about the quality of particular services sent to local governments in text messages is imagined to create an infrastructure of accountability that will feed information to the national government and ultimately to the donors.

As this example indicates, information infrastructure design is guided by the aspiration to enable people and organizations to collaborate in ways that bypass concerns with time and space. Real-time infrastructures are envisioned in which anyone can access data from anywhere at any time. Indeed, the sociologist Manuel Castells (1997) has conceptualized modern infrastructure as a "space of flows." Yet the metaphor of information as a liquid that flows freely in any direction has several limitations (Moser and Law 2006). For one thing, information generally doesn't flow quite so freely; infrastructure contains classificatory as well as material boundaries. Information doesn't suddenly flow out of the categories and classification it is contained in, nor does it jump from one technological platform to another all by itself. At the same time, no infrastructure boundary is entirely static. And the fact that information is constrained in various ways doesn't mean that it is easy to control, or that it is always effective. (See chapter 6.)

Meanwhile, science and technology studies have shown that infrastructure is not an exclusively technological domain. On the contrary, infrastructures, whether in the form of roads (Dalakoglou 2010; Harvey 2010; Nielsen 2012) or in the form of databases (Bowker 2005), are entwined with cultural habits, with social and economical organization, with professional and personal identities, and with many other things. Since the notion of infrastructure as underlying basis no longer suffices, diverse scholars have begun to re-conceptualize infrastructures as complex systems, socio-technical networks, and heterogeneous assemblages. (See, e.g., Egyedi and Mehos 2012.)

Still, this is not a majority position. Indeed, studies of infrastructure continue to be mainly based on the assumption that infrastructures are primarily technological, and that they provide a basic structure through which, or on top of which, other (organizational) activities occur. In their 2007 report "Understanding infrastructure: Dynamics, tensions, and design," Paul Edwards, Steven Jackson, Geoffrey Bowker, and Cory Knobel spend considerable energy countering this reductive tendency. They emphasize that the view of cyberinfrastructure as simply a "machine to be built" or a "technical system to be designed" implies a downplaying of "the importance of social, institutional, organizational, legal, cultural, and other non-technical problems developers always face."

Failure to recognize the heterogeneity of information infrastructures and the practical challenges this variability entails may be organizationally and economically costly. It is also analytically costly. Consequently, we will do well to heed Susan Leigh Star and Karen Ruhleder's (1996) advice never to assume that infrastructure are fixed all at once or globally; that they are never changed only "from above." Although change processes are often initiated by powerful organizations, actual change always occurs through negotiations and alignments that implicate many more actors in much more complex and surprising ways.

On the one hand, then, infrastructures are often conceived "from the top down," that is, by managers or politicians, basically as machines to be built and implemented in order to solve various organizational and collaborative problems. On the other hand, they must in reality always be tinkered with—"from the bottom up"—to become workable. In view of this discrepancy between conceptualization and practice, it is little wonder that infrastructures often become objects of perplexity and controversy. Edwards et al. summarize this situation well, suggesting that infrastructure tends to be seen simultaneously as an "all-encompassing solution and an omnipresent problem, indispensable yet unsatisfactory, always already there; yet always an unfinished work in progress" (2009, 365). The next section considers how researchers at the intersection of STS and information studies research have grappled with the empirical and conceptual issues pertaining to infrastructures.

From Information Systems to Information Infrastructures

In the field of information systems, researchers gradually began to realize that they lacked appropriate tools for analyzing the distributed, networked nature of modern technologically mediated work. Especially in the 1980s, computers and electronic networks became increasingly pervasive features of organizational life. This proliferation of information systems created challenges relating to questions such as how to align networks, applications, and databases with each other and with diverse distributed work practices. (See Edwards 2010.)

The notion of *information infrastructure* became central to conceptualizing this situation. It enabled researchers to question standard assumptions about both technologies and organizations. Among these assumptions were taken-for-granted ideas about the nature of information and its relation to such diverse issues as collaboration, knowledge sharing, division of labor, control, and strategy. In this context, the notion of information infrastructure became part of a project that was simultaneously empirical, conceptual, and critical.

According to Claudio Ciborra and his collaborators in *From Control to Drift: The Dynamics of Corporate Information Infrastructures* (2000), existing theories of infrastructure had largely outlived their usefulness by the 1990s. These were approaches that had theorized the corporation as an internally coherent and bounded entity. In the estimation of Ciborra et al., most of these studies had addressed infrastructures, and the technologies and standards they comprised, as if they existed in an organizational and a social vacuum. Although this assumption might have been suitable for earlier "mechanical organizations," Ciborra et al. suggested, it was certainly inadequate in the present moment. Organizational actors needed to act strategically in situations characterized by the interdependence of the corporation and an increasingly fluid environment:

> [T]he complex process of 'wiring the corporation' cannot be understood, let alone managed, by applying approaches that were effective for mechanical organizations, and assembly-line type of technologies and processes. Information systems are often

analyzed and designed as highly decomposable; bureaucracies are also decomposed along historical lines. Extant information infrastructures, we claim, are deployed within ramified webs of externalities and interdependencies. (2)

This line of criticism was used in support of studies of a changing, technologically mediated corporate situation in which organizations would increasingly collaborate across geographical distance by means of information and communication technologies. Through these suggestions, a problem space was defined that was akin to the one Edwards et al. identified in their introduction:

How to integrate or replace existing systems; how to handle local divergence from a desired norm; including how (or whether) to keep users from working around or outside the system; and what is gained and lost in the transition to a new infrastructure. (2009, 366)

As Edwards et al. noted, attempts to address such questions take us "below the surface level of goals and implementation into the murky depths of tension, failure and compromise where conflicting purposes and political differences meet technical details" (366).

For Ciborra et al., the novelty of the situation had to do with the complexity and scale of new technological platforms. In addition, it related to the inability of standard organizational approaches to understand and sensibly guide the design and use of these technologies. The term *information infrastructures* signaled precisely the proposed shift in analytical approach: it would replace the predominant focus on information *systems*, broadly characterized by its *a priori* separation of technical, social, organizational, and political issues. In contrast, information infrastructures were supposed to make it possible to develop better and more realistic analyses of changing infrastructures. It would also help address the question of how organizations themselves changed—or drifted—along with their infrastructures. These were studies, then, that aimed to take seriously the dynamics and the heterogeneity of IT in modern corporations:

Corporate information infrastructures are puzzles, or better collages, and so are the design, and implementation processes that lead to their construction and operation. They are embedded in larger, contextual puzzles and collages. Interdependence, intricacy, and interweaving of people, systems, and processes are the culture bed of infrastructure. (Ciborra and Hanseth 2000, 2)

Thus, the argument that managers should think about technology as embedded in and interdependent with organizations was meant as both analytical innovation and practical provocation. Especially provocative were the organizational implications of following this line of thought, since it entailed abandoning the belief in organizational control that was central to much managerial thinking. By arguing that the dynamic and interrelated nature of technology and organizational development should be acknowledged, then, these authors threw sand into the control-oriented machinery of conventional management theories about both organizational process and technological facilitation.

As has been noted, fluid and malleable connections between technologies, organizations, and people were central to the agenda of information-infrastructure studies. The fluidity provided an important argument for shifting managerial orientations from control to drift. However, there was yet another reason why managerial control of infrastructures was seen by Ciborra and colleagues as unlikely to succeed. The linchpin of their argument was that, although infrastructures are continuously drifting, the direction of such movement is significantly determined by an underlying "installed base" of standards and technical configurations that are not easy to change. Newly introduced technologies are therefore never "simply" new; they never stand alone, because they must always be connected with existing infrastructures. This is why radical changes that managers or technologists hope to accomplish by way of infrastructure are often somewhat unrealistic; they fail to take into account that existing infrastructures retain their own ingrained inertia and sense of direction. Thus, even as this argument challenges technological determinism in a strong form, determinism is retained in a softened version. After all, organizations seem to drift *on top of* infrastructures whose installed bases make them largely inert.

The Politics of Infrastructure

Many of the points we discussed in the previous section were inspired by analytical work in science and technology studies (e.g., Akrich 1993; Bishop and Star 1996; Latour 1996). STS provided inspiration through, among other things, their attentiveness to the heterogeneity of the actors that make up technological and scientific networks and practices. (See, e.g., Bowker 1994a.) In STS this analytical tenet had been well established by such analytical approaches as actor-network theory, the social construction of technology, and symbolic interactionism, and by historians of technological systems. In addition, Susan Leigh Star (1995a), Lucy Suchman (1987), and Diana Forsythe (2001) helped to bridge the gap between information systems research and STS and to bring the agendas and concerns of those fields to bear on each other. They developed agendas that focused on the politics of technology (see Berg 1998) and on the politics of classifications and standards embedded in technological work. That body of work also emphasized the invisible social and organizational work on which smoothly functioning systems and infrastructures depend. In conjunction, these studies demonstrated that infrastructures invariably both hinder and facilitate a variety of aims and agendas. They made abundantly clear that infrastructures had innumerable effects—some small, some large—that shaped organizational work and politics. (See, e.g., Blomberg, Suchman, and Trigg 1996; Star and Strauss 1999; Vaughan 1996.) In contrast to scholars who looked for ways of improving the understanding and application of large-scale information technological systems, for STS researchers the central question was "What made infrastructure in the first place?"

Sensitized by an interest in how politics come to be embedded in information infrastructures, researchers began looking in more detail at infrastructures' distributed and dynamic evolution. Infrastructures gradually

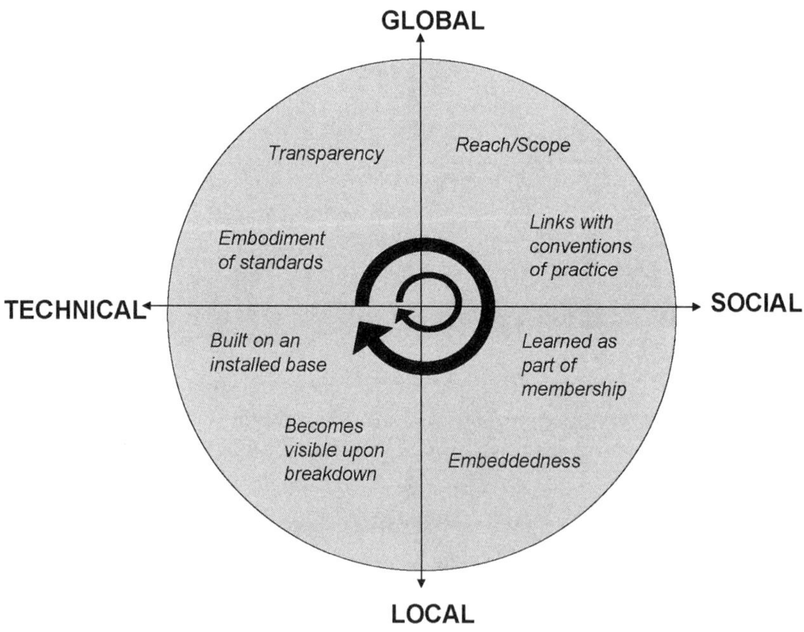

Figure 1.1
Cyberinfrastructure as distributions along technical/social and global/local axes. Reprinted with permission from Florence Millerand, Karen Baker, and Geoffrey Bowker.

came to be seen as "political machines" (Barry 2001) embedding particular scripts that specified preferred users and patterns of use (Akrich 1992; Oudshoorn and Pinch 2003; Woolgar 1991). In this tradition, a growing number of studies have pointed to multiple dimensions and to both intended and unintended effects of growing, using, managing, and working with infrastructures.

Sundeep Sahay, Eric Monteiro, and Margunn Aanestad (2009), for example, studied health-care information technology in India and showed how assumptions about the integration of information systems that are taken for granted as a "good" in a Western organization is interpreted very differently in "other contexts." In a related vein, though in a completely different setting, Brit Ross Winthereik and Signe Vikkelsø (2005) highlighted the efforts required to integrate the work of physicians with new hospital infrastructures by focusing on negotiations and alignment of data. These studies began to open the black box of infrastructures, both empirically and analytically. (See also Aanestad 2002; Pirnejad, Bal, and Berg 2008; Vikkelsø 2005.) In doing so, they broke with the assumption of linearity inherent in notions such as information flows and chains, and paved the way for a view of infrastructure as recursive or fractal (Jensen 2007a).

Scales and Ontologies of Infrastructure

The bodies of research that we have discussed so far focus on the fluidity of information and on the politics of infrastructures. Because such studies explore the socially embedded nature of information infrastructure, the preferred research strategy has been to conduct detailed empirical studies of particular organizational sites and arrangements.

This research strategy, however, has recently come under critical scrutiny (Beaulieu, Scharnhorst, and Wouters 2007; Wyatt and Balmer 2007). Neil Pollock and Robin Williams (2010) have suggested that single-site research is inadequate if one wants to understand the slowly evolving, long-term development of larger, deeper, and increasingly global infrastructures. Pollock and Williams argue that some of the most important information infrastructures (their chief example is SAP, which stands for "system analysis and program development") span very diverse settings and evolve over long periods. Such evolutionary processes, they argue, cannot be understood by studying any particular setting, no matter how detailed the study. Consequently, they advocate a methodological perspective that encompasses both short-term dynamics and the long-term evolution of infrastructures. Following Ribes and Finholt (2009), Pollock and William present their interest in "bridging multiple temporal scales" (2010, 538) as akin to the historical studies of the *longue durée* advocated by the French *Annales* school in history, which focused on very-long-term transformations of trade, politics, infrastructure, and social habits across whole regions over decades or centuries. (See Braudel 1976.)

Pollock and Williams' response to the problem of how to analyze complex systems distributed among multiple settings is *strategic ethnography*. It is defined by a focus on technological development through the product cycle and an interest in the specificities and workarounds of contextual use of technology in specific organizational settings. Pollock and Williams characterize this approach as a study of "biographies of artifacts." (Cf. Daston 2000.)

As Pollock and Williams see it, the ethnography of infrastructure must be strategic in its choice of field sites. Now, of course, ethnography is never freed from analytical motive (see Jensen 2007b); in this general sense, all ethnography can be said to embody strategies. But whereas most anthropologists would grant their informants significant powers to reorient what counts as strategic relevance or location (see, e.g., Ho 2009; Marcus 2005), this choice is made in advance by strategic ethnography. Needless to say, this circumscribes the field of study and forecloses some of its openness. The emphasis on strategic decisions is, however, directly connected to Pollock and Williams' broader interests and agendas. Referring to the studies conducted by Ciborra and colleagues, Pollock and Williams worry about the "pessimistic" implications of focusing on "drift," which, they suggest, leads to analyses that "run the risk of offering a fatalistic account of the necessity of failure" (2010, 539). However, they argue that their own methodology can "guide policy and practice" (522) in more constructive ways.

Specifically, Pollock and Williams seek a framework that is capable of taking into account both the temporal and the spatial dimension of changing infrastructures. Thus, they propose the use of longitudinal studies across a variety of sites, and they advocate a combination of analyses that center on the "micro-context" of individual actors or workplaces and studies of institutional macro settings. This methodology offers the researcher a perspective from which to look across sites of infrastructure while also tracing their temporal development. If a synchronic gap is identified, it can be filled by adding more sites. If a diachronic gap is identified, it can be filled by following interactions over longer periods. In both a spatial and a temporal sense, then, this is an approach whose other is a study of a single-site case over a relatively limited period.

Pollock and Williams distance themselves from the "modernist dream" that views information infrastructure "as a universal solution that can somehow anticipate and cater for the entire range of organizational contexts and practices" (2010, 533). But although they are skeptical toward modernist views of technological solutions, their methodology and their epistemology seem distinctly modernist. For one thing, their approach is based on the modernist assumption, simultaneously empiricist and holistic, that infrastructures can *in principle* be comprehensively characterized by adding together enough detail from enough sites. In that sense, their approach can be contrasted with the nonmodernist (Latour 1993) or post-plural (Strathern 1992a) ethnography that inspire much of what follows. As we shall see, such ethnography is based on a fractal, recursive imagery of infrastructure that is sharply at odds with the notion that infrastructure can be defined in terms of a substratum of technology underlying and perhaps determining social situations that take place "on top of" it.

Additionally, Pollock and Williams' approach assumes a particular position of epistemic authority for social science analysis. In contrast, we see social scientific intervention in design as an effect of alliances, betrayals, and negotiations. This means that the interventionist *ambitions* of one's own analyses are only of limited consequence. (See Jensen 2007b.) These observations, finally, also indicate why Pollock and Williams' critique of drift fails to hit the mark. Interpreting the concept of drift as based on a normative rather than a descriptive stance, they take Ciborra et al. to task for a "pessimism" that is in fact a descriptive indicator of their underlying disagreements about the constitutive features and operations of infrastructure.

As the social anthropologist Marilyn Strathern suggested, postplural anthropology dispenses with the assumption that ethnography must be premised on, or working toward, a "whole picture"—a whole infrastructure, for example—that can be grasped analytically. The reason is that, for any analytical endeavor, "the perception and filling of a gap lead to awareness of the gaps" (Strathern 1989, 63). The more empirical or analytical "gaps" are filled through study, the more gaps become visible in turn. Because sites don't "add up," the multiplication of infrastructural sites of study, proposed by Pollock and Williams, will not automatically help to create a more comprehensive picture. Instead, each site, "small" or "large," generates both knowledge and gaps of knowledge, as do the sites in combination. This imagery is fractal because it pictures the complexity of infrastructure as scale invariant, with complexity replicating across scales. Another implication, discussed in the following chapter, is that researchers' ways of making knowledge about infrastructures become recursively implicated with infrastructures themselves. (Cf. Dormans and Kok 2010.)

Pollock and Williams aren't alone in questioning the adequacy of single-site ethnography in the context of infrastructure. The historian of technology Paul Edwards, too, finds local studies of technology development to be methodologically insufficient for understanding how information infrastructures tie together multiple organizations, forms of work, modes of collaboration, and political and economic agendas in increasingly global networks.

For Edwards, infrastructures are central actors of modernity: everyone lives with (and against) them. Indeed, Edwards suggests that proficiency in navigating infrastructure is an important "ethnographic indicator" of cultural and social membership. Thus, for example, "societies whose infrastructures differ greatly from our own seem more exotic than those whose infrastructures are similar to our own" (Edwards 2003, 189). This argument opens up infrastructure to a very broad set of social theoretical and ethnographic concerns.

For example, Edwards argues that infrastructures are "largely responsible for the sense of stability of life in the developed world, the feeling that things work" (2003, 188). This expansive view sees infrastructures as much more than a matter of linking technologies. Rather than allowing organizations and people to do things, infrastructures also define what can be known. Thus, they also shape imaginations of what can be done. Accordingly, infrastructures literally change our worlds and our understandings. We can therefore extrapolate from Edwards' characterizations to argue that infrastructures must be studied for their ontological, world-shaping implications. Emerging infrastructures articulate practical ontologies (Jensen 2010a), the consequences of which must be elicited empirically and conceptually.

This argument can be elucidated further by juxtaposition with Bruno Latour's discussion of "the modern settlement" (1993). In *We Have Never Been Modern,* Bruno Latour analyzed the double movement of modernity whereby on the one hand "moderns" conceptually purify nature from society while on the other hand they never cease to create new mixtures of nature and society. (See, e.g., Latour 1993, 2004.) As we have seen, in Edwards' view, moderns, rather than operating in a natural world, literally live in infrastructures, which function as their invisible social environments. Infrastructures have, then, become social and organizational "second nature" (Bowker 1995) for moderns. Practical ontology, in this context, might be described as the study of how infrastructures become second nature, and what kinds of second nature they do become. And just as analyses were difficult to carry out in the case of first nature, by definition understood as natural and thus almost impossible to question, so it is hard to gain either analytical or ethnographical distance from infrastructures, which, also by definition, tend to become invisible, to "fade into the woodwork" (Bowker and Star 1999, 34) as second nature. How, then, to get infrastructure and their effects into analytical view?

Contours of Infrastructure

The approaches we have discussed share the view that infrastructures are fundamentally heterogeneous, not just in their technological components, not just in the perspectives people have on them, but also in their ontologies. For infrastructure, design and implementation require not only the installation of technical standards but also the creation of new things in the world: new ways of knowing, new kinds of people, and new material objects. For example, as we will discuss in chapter 3, aid technologists construct not only new databases and information repositories but also new figurations of users, such as "users on the ground" in developing countries and "concerned citizens" at home. And, as we will discuss in chapter 4, infrastructure development entails the shaping of new informational ontologies and spaces of habitation that transform the political agency of users and of organizations.

Studying infrastructure as emerging, practical ontologies necessarily implies facing certain methodological challenges. As we have seen, Pollock and Williams and Edwards all question whether single-site studies are appropriate for studying the long-term, dispersed effects of infrastructure. Yet, as we have begun to suggest, the solution proposed by Pollock and Williams runs up against a series of recursive empirical and conceptual challenges when studying emerging aid infrastructures of accountability and transparency.

What are the issues? First, even though the biography-of-artifacts approach acknowledges that, "in the early stages," "development processes may be relatively open" (Pollock and Williams 2010, 546), this still assumes a degree of coherence, stability, and materialization of infrastructure. Without such stability and coherence, it will not be possible to recognize particular socio-technical assemblages as infrastructure. A researcher who assumes that infrastructure operates by linking together many different organizational and inter-organizational levels (cf. Schmidt and Werle 1998) is invariably led to posit some kind of coherent infrastructural whole.

Yet in several of the cases discussed below we encounter emergent infrastructures that are, in fact, rather wildly incoherent and only partially materialized, and that continue to pull in different directions. It is uncertain whether there is, in fact, *a* biography to describe in such cases, or whether instead there are multiple competing versions of infrastructural "identity." What we describe are, thus, series of different visions, techniques, aims, and activities, leading to various, only partly overlapping versions of infrastructure. Another way of saying this is that because distributions of agency and power are at stake in emergent aid information infrastructures, there are always many competing "wholes" in play. And indeed these various visions and versions are instigators of the processes of infrastructure making that we study. This is why our study turns to a series of concrete but divergent infrastructural initiatives, even though our ambition is also to understand how the monitoring movement aims to create transparency and accountability globally.

The import of this observation is that localization must still be taken seriously, though not in the guise of the "single local case," which Edwards and Pollock and Williams have rightfully criticized. For consider how strange the terminology of the local is in the context of global aid information infrastructure. In situations in which the question of how practices might be connected is at stake, the local ceases to operate as expected. The local and the global flicker: changes in scale and transformations in power become possible. And even when infrastructure does connect distributed practices, this never happens seamlessly, or in all dimensions.

In this situation, it isn't possible to make sense of emergent infrastructures either by conducting single-site studies or by making encompassing biographies of how they perform globally. Single case studies are insufficient because aid infrastructures are elements of a global vision even if they never materialize in accordance with such a vision. Yet even so, the tentacles of aid information infrastructures are already simultaneously moving in many different directions. A "configurational politics" of infrastructure (Sahay, Monteiro, and Aanestad 2009) is already operational, and an unstable "translation terrain" (Pollock and Williams 2010, 538) is in place.

The existence of this translational terrain is methodologically consequential. It means that focusing exclusively on the biography of some high-level infrastructure project, initiated by powerful aid organizations (chapter 4) or audit offices (chapter 6), would leave us blind to the fact that infrastructures of accountability are already part of terrains elsewhere.

It is precisely because we are following emergent infrastructures that we can assume neither their stability nor their ability to connect. Constantly trailing our informants in these processes, we need to come to terms with infrastructure as an expansive and open-ended but also fundamentally vague ontological entity (Winthereik and Verran 2012).

In this situation, vagueness is not a weakness of infrastructures; rather, it is a condition for their emergence and workability. It indexes that the contours of infrastructure are continuously drawn by a multitude of actors who have different ideas about the configurations they aim to create. Hence, our explorations of emergent infrastructures for monitoring and evaluation assume that such tools and technologies are never independent of the actors and organizations that promote, maintain, and use them. Accordingly, they are also unlikely to be "the same" even in cases in which they are promoted with visions and metaphors that seem superficially akin.

Development Histories

In their 1996 book *Doctrines of Development*, Michael Cowen and Robert Shenton offers a genealogy of the concept of development that harkens back to colonial times. They note that development as an idea was originally invoked not to resolve issues in "developing countries" but rather to deal with issues at home. Development was "invented, amidst the throes of early industrial capitalism in Europe" to "construct order out of the social disorders of rapid urban migration, poverty and unemployment" (5). Cowen and Shenton note that development now "seems to defy definition, although not for a want of definitions on offer" (3). Centrally, they argue that the notion of development suffers from inherent contradictions. Specifically, development conflates two quite different things. On the one hand, "development usually presents an image of something created anew, or improved, or renewed, or of the unfolding of potential which has the capacity to exist but which presently does not do so" (viii). On the other hand, "to intend to develop does not necessarily mean that development will result from any particular action undertaken in the name of development" (viii). Trouble occurs, therefore, when development as an immanent process is confused with development as an intended aim. For Cowen and Shenton, an intention to develop turns into a "doctrine" when it becomes "an expression of state policy" (viii). In such cases, development entails "trusteeship," an idea according to which "those who took themselves to be developed could act to determine the process of development for those who were deemed less-developed" (4). By the late twentieth century, however, this "logical sleight of hand" had ceased to convince. "Trusteeship stands condemned as Eurocentrism, an imperial vestige of the post-1945 attempt to improve living conditions of poor colonies and poor nations through state administration." (4)

Cowen and Shenton trace the history of development back to early industrialism. More often, however, the history of development is seen to begin after World War II. The Bretton Woods institutions (the International Monetary Fund and the World Bank) had been established in 1944, the Food and Agriculture Organization in 1945 (Gardner and Lewis 1996, 8–9). In this version of the history of development, "development first emerged as a transnational mission in the 1950s" (Finnemore 1997, 203) in order to further economic growth. One notable promoter of this agenda was President Harry Truman. The development project was also inseparable from the context of the Cold War, which "made aid politically attractive for governments anxious to stem the flow of communism in the South" (Gardner and Lewis 1996, 10). In tandem with decolonization (Cooper and Packard 1997a, 7), "aid programmes expanded rapidly" (Gardner and Lewis 1996, 10). Indeed, donor countries began competing to provide assistance to developing countries.

Broadly aligning with Cowen and Shenton's analysis of the doctrine of development, the anthropologist Richard Rottenburg also argues that development is a continuation of colonialism (2009). In the following chapters, we interrogate this claim by attending to information infrastructures created to support a purportedly new mode of development assistance based on transparency, mutual accountability, and partnership. Thus, emerging information infrastructures and partnership are our entry points for examining current practical ontologies of technologically mediated development aid.

The 1960s saw the establishment of bilateral agencies such as the US Agency for International Development (1961), the British Overseas Development Administration (1964), and the Danish International Development Agency (1962). Just like the multilateral organizations, bilateral operations were heavily influenced by modernization theory as it was formulated within political science and economics. According to the classic formulation of this theory by the economist Walt Rostow (1960), who worked with a model based on stages of growth, "traditional society is poor, irrational and rural," but it can—and will—move forward "based on technology and high levels of investment" (Gardner and Lewis 1996, 12–13).

Gardner and Lewis emphasize that modernization theories are inherently optimistic in their assumption that all countries will "eventually experience economic growth" (13). In addition, such approaches present modernization as a "relatively easy process," and "enduring underdevelopment is explained in terms of 'obstacles'" (13).

In the 1970s the focus of development was redefined and reoriented. Modernization retained its strong appeal but was gradually complemented by a poverty orientation, an agenda that Robert McNamara promoted as president of the World Bank (Finnemore 1997, 204). This "basic needs" approach stressed the importance of combating poverty rather than promoting industrialization and modernization, a re-orientation that saw development cooperation as a moral achievement. The report *Partners in Development* (Pearson 1969)—the result of McNamara's request to have Lester Pearson form a commission to study the consequences of development, assess the results, and propose policies—included consultations with governments in developing countries all over the world. The final report argued that development could have positive implications for building "the world community." While emphasizing the role of development for trade and overpopulation, the report nevertheless contrasted starkly with the "modernization" perspective, which largely left "the aid recipient" out of the picture. By introducing the idea of developing countries as "partners," it attempted to humanize development and to show respect for countries that had previously been considered as residing at a lower stage of evolution. Even so, the countries in question were still considered in need of development and modernization.

The 1970s also saw an expansion of development themes and methods, as indicated by interest in community participation, gender issues, sustainability, and partnership (Cooper and Packard 1997a, 24). The word *partnership* came to refer to various kinds of cooperation between states. However, in the 1970s development was also popularized through greater involvement by grassroots movements. Thus, participation in development activities was promoted as part of a solidarity agenda.

In connection with the debt crises of the 1980s, the development agenda took yet another turn. Once again, focusing on economic growth and on restraining public expenditures came to be seen as crucial for development (Gardner and Lewis 1996, 6). Measures to ensure restraint in public expenditures were implemented through "structural adjustment programmes" that entailed a dual focus on "market discipline" and "good governance"—"the imposition of political conditions . . . on the provision of aid" (Cooper and Packard 1997a, 22). At the time, structural adjustment was promoted as a way to help poor countries out of their debt. However, they came to be seen as spectacular failures both by market liberalizers and by critics of development (Easterly 2002; Ferguson 2006a; Stiglitz 2002).

In the 2000s, structural adjustment was complemented by "poverty reduction strategies" intended to be more flexible and to be tailored to the needs of specific countries. Critics, however, argued that these strategies, whether intentionally or not, supported legitimization of the "continuing donor emphasis on policies of trade liberalization" and economic discipline

(Mosse 2005a, 3), so that they are most accurately described as "'soft' or 'processual' conditionalities" (Gould 2005, 63). (See chapter 2 below.)

Today it is widely observed that development aid has been permanently "neoliberalized"—that is, denationalized and decentralized—in such a way as to "diversify further sources of power and influence, via a proliferation of organizations and intermediary networks" (Lewis and Mosse 2006, 2). Thus, in macro-political terminology, we are witnessing the emergence of a new "global aid architecture" (Mosse 2005a, 3) that relies on donor collaboration "to assist governments make their own overall strategies" and "sector-wide approaches" (3) based on "basket funds" and "direct budgetary support" (Gould 2005, 62) rather than on individual projects. While this has opened up many new avenues for business to enter development in partnership with non-governmental organizations, bilateral organizations, and multilateral organizations, it has also meant that "advocates of participatory alternatives are no longer confined to the grass roots, but promote participation at the macro level"—for example, "through civic engagement," "secondary stakeholder participation and partnership as a mode of donor-recipient relations, or by contributing to networks of global civic activism" (Lewis and Mosse 2006, 6). We are thus witness to a situation in which various kinds of capacity-building efforts on the ground coexist with "securing institutional buy-in to market-based patterns of liberal capitalism and governance structures" (8–9). No wonder the "official narratives of development institutions are concerned with attempts to 'square the circle'" (8). One way they attempt to do so is by experimenting with new aid infrastructures that aim to facilitate accountability and transparency between partners.

Accountability, Information, and Transparency

Infrastructure and information were always elements of development. On page 1 of his 1993 book *The Imperial Archive*, Thomas Richards notes that as early as the 1860s Britain was "full of fantasies about an empire united not by force but by information." They were fantasies because "people were painfully aware of the gaps in their knowledge" (3). Though administrators did their best to fill these gaps, they weren't able to manage and control the Empire, so they "could do little other than collect and collate information" (3). Colonial information refused to add up to a whole picture. Later in the nineteenth century, many people gave up on the idea that amassing information necessarily added up to creating knowledge. Already, the compilation of seemingly endless streams of data created a "bottomless pit"

(15). The very notion of information, with "its contemporary overtones of scattered disjunct fragments of fact" (5), is indicative of the situation.

In mainstream post-1945 development, infrastructure was typically conceived in the classical sense of Edwards' "fixed facilities"—necessary components for the achievement of economic growth. For example, "inadequate infrastructure" might be explained in terms of "lack of capital, weak or corrupt management and lack of local expertise" (Gardner and Lewis 1996, 14). Thus, in principle, infrastructural issues could be resolved easily: simply bring in capital and expertise to build roads and bridges, train local people, and support good government. Though the terminology was different, notions of transparency and accountability were in play in such processes. Good governance opposed bad governance, which meant corrupt government. The creation of "transparent" bureaucracies, economically accountable and thus modern, was intimately related to efforts to build infrastructure.

We show in this book that infrastructure and information have been coupled with development and connected to issues of accountability transparency, and partnership in ways that reconfigure the ontologies of development aid. The 2005 Paris Declaration on Aid Effectiveness is indicative of this situation. These keywords (Williams 1981), alone or in combination, are found in any current discussion of development. Thus, it is particularly striking that none of the books we have drawn upon in outlining the recent history of development (published, as they were, between 1996 and 2006) contain any index references to either infrastructure, partnership, or transparency. The sole reference to the currently omnipresent term *accountability* relates not to development organizations but to anthropologists studying their practices (Gardner and Lewis 1996, 135).[1] Gardner and Lewis note only that one latter-day strategy for the improvement of infrastructure "might be the introduction of information technology to local institutions, or the training of personnel to use new technology" (1996, 14).

Yet we are now witnessing concerted efforts to create global information infrastructures for development. Richards' (1993) "imperial archive" centered on collecting information is being rapidly updated with new technological tools, from databases to geocoding applications. As we shall discuss, however, this doesn't make its vision for the global development archive any less utopian.

In the eyes of designers, NGOs, and policy makers, the outcome of these infrastructural initiatives is to make aid development more transparent and thus more accountable and equal. In these processes, aid infrastructures undergo transmutations akin to those we traced in the context of information studies. Their ontologies are morphing: they disperse and become fluid, they spread everywhere, and they impinge on the practices and aspirations of new actors.

Aid infrastructure development promises heightened accountability between donors and recipients re-described as partners. In one sense, emerging aid information infrastructures can be seen as updated versions of Richards' imperial archive: "the collectively imagined junction of all that was known or knowable, a fantastic representation of an epistemological master pattern" (1993, 11). Though such infrastructures might provide an Eldorado for people seeking answers to epistemic puzzles, in another sense infrastructures are also effects of work. Studying both this work and these puzzles through the lens of emerging practical ontologies allows us to focus on how infrastructures generate new questions, new concepts, new users, and new politics—that is, new worlds of aid.

Anthropological Positions

As is the case for infrastructure, the study of development aid is not the prerogative of any particular discipline. If development aid is very broadly related to the efforts that Western countries have put into transforming the social and technological conditions of non-Western countries for the better, the reasons given for creating such improvements and modernizations, and the efforts taken to create them, have been analyzed by multiple disciplines and from many perspectives. Historical studies of the genealogy of the development concept are available (e.g., Cowen and Shenton 1996). Cooper and Packard's book *International Development and the Social Sciences* (1997b) contains contributions from anthropology, history, and political science. Cultural geographers and postcolonial theorists are engaged in studies of development. STS is hesitatingly entering the fray. (See, e.g., Anderson 2002, Burrell 2011; Langwick 2007; Verran 2002.[2]) Development economics remains by far the most influential discipline dealing with the topic. However, given the different forms of analytical attention paid to development from this wealth of disciplines, it is no wonder that *development* seems a fundamentally unstable term (Edelman and Haugerud 2005, 1).

David Mosse and David Lewis (2006) propose a typology according to which anthropological engagement with development can be characterized as instrumental (applied, managerial, bureaucratic), as ideological (in support of participation, "bottom-up," or "alternative" development), or as deconstructive (criticizing and undermining the notion of development *tout court*).

From the "ideological" side, the notion of development has long been bound up with a reflexive critique of the politics of ethnographic knowledge making (Asad 1973). The days when a prominent anthropologist could end an exposition of social anthropology for the general public with a chapter on "applied anthropology" (Evans-Pritchard 1951, 109–131), arguing for the value of ethnography for administration and colonial government, are long gone.[3] A critical and reflexive view of the intimate relations between anthropology and colonial power has emerged.

A critical narrative that has long been popular in the anthropology of development describes how aid practitioners, blinded by ideology, propagate irrelevant advice or enforce harmful initiatives. Thus, anthropologists have repeatedly criticized aid interventions for failing to take into account—or actively ignoring—existing social, cultural, and political contexts. The critical implication of this diagnosis is that decontextualized projects tend to run roughshod over the interests and livelihoods of affected people. (See, e.g., Goldman 2005; Scott 1998.) This line of criticism has also extended into a more fundamental critique (Lewis and Mosse's "deconstruction") of the depoliticizing effects that development organizations bring along. (See, e.g., Escobar 1995; Ferguson 1990.) Thus, contentious social and political problems are translated into problems to be solved by technical means that can be provided by aid organizations. And some of those translators are applied anthropologists (Lewis and Mosse's "instrumentalists").

In a 1997 article titled "Anthropology and its evil twin," James Ferguson describes the long-standing and tense relationship between "pure" and "applied" anthropology. Early anthropological efforts were often directly engaged in constructing the colonial archive. In Rudyard Kipling's *Kim*, the figure of Colonel Creighton is noteworthy for his dual role as head of the ethnographic survey and chief of the Indian secret service. Thomas Richards calls him "perhaps the most cybernetic character in Victorian literature" (1993, 39).

Perhaps as a consequence of the real and imagined existence of such anthropological pawns of colonialism, the field of anthropology came to establish increasingly clear lines of demarcation between pure anthropology and its messy application. Even so, Ferguson argues that this demarcation is itself largely imaginary. On the one hand, academic theories have been hugely influential in practical affairs of development, though "they have not always been applied and used in the way that their academic originators might have wished" (1997, 151). On the other hand, applied anthropological research, as exemplified by the Rhodes-Livingstone Institute, has also contributed to shaping the agenda of academic anthropology.

In tandem with the changing agendas of development aid since the 1960s, a "new terrain for academic inquiry" emerged (Cooper and Packard 1997a, 13). It also created openings for anthropologists. In particular, Ferguson argues, the 1970s' orientation toward "basic needs" and the "'soft,' 'social' side of development policy" created a "burst of anthropological interest in development," and, indeed, created the subfield of "development anthropology" (1997, 164). It created space for both "instrumentalists" willing to work for development organizations and "ideologists" engaging in critical analysis of its implications.

Development anthropology, however, had a hard time. Within development agencies, anthropologists were disadvantaged because of their interest in the complexity of development problems and their tendency to "serve up post-hoc criticism of failed projects"—an emphasis, Ferguson says, that "quickly became a kind of anthropological specialty" (165). Meanwhile, within academic anthropology, development anthropology acquired a reputation as a rather low-prestige subfield.

Ferguson's analysis of the troubled relationship between anthropology and development focuses on the particular ethnographic obsession with "the local." He notices that, whereas in principle "the local" could be anywhere, in reality the privileged location of ethnography is usually "not only 'local,' but muddy, tropical, disease-infested" (169). Indeed, he argues that a version of development doctrine has also facilitated the anthropological distinction between appropriate and inappropriate sites of study: those places "that have not experienced development are most anthropological; those that are 'developed' are least" (169). Accordingly, the archetypical field position of the critical anthropologist of development is among the people at the receiving end of aid initiatives, and thus is in an excellent "insider" position from which to identify and criticize unforeseen and paradoxical consequences of aid initiatives (the anthropological specialty of *post hoc* criticism referred to by Ferguson).

Inside and Outside Development

From the point of view of our project, the critical outsider position is untenable. For reasons that will be specified further in chapter 2, we have had access to no privileged "local" site. The only time we went to a village, we traveled with aid consultants, focusing on their monitoring work. We left when they left.

Yet the untenability also has a more principled character. In fact, we doubt that any privileged local site exists in our case. Insofar as we wanted to study the *global* information infrastructure of development, our field was, as it were, automatically dispersed. By definition, aid infrastructures move across contexts. Indeed, in our case, as in many others, they were explicitly created to "link" the "local" with the "global." It is clear that the study of infrastructure locates us in a landscape that cannot be defined by a single "inside" locality strictly demarcated against an "outside" globality, for emerging infrastructures span multiple insides and outsides. No matter where we located ourselves, we would create both an inside position and an outside position. (See Mosse 2005b; Yarrow and Venkatesan 2012.)

As a consequence, also, we have difficulties finding ourselves in Lewis and Mosse's typology. We are certainly not instrumentalists, though we have studied NGOs. We are definitely not ideological populists or development critics, though we are quite critical of certain implications of global infrastructure development. However, we also do not view ourselves as deconstructivists, though we are inspired by James Ferguson's *Anti-Politics Machine* and similar works. Rather than critical or applied, our view of infrastructures as giving rise to emergent practical ontologies of development aid might be designated as symmetrical and acritical.

Our ethnographic position relative to the anthropology of development is curious both in analytical terms and in methodological terms. We begin on the putative "inside" of applied development research initiatives that aim to build more transparent and accountable infrastructures, but our aim is to loosen the very distinction between insides and outsides. Accordingly, our central interlocutors from STS and anthropology are also researchers who position themselves as neither quite one nor the other.

In view of this position, it is worthwhile to briefly consider how the problems of development look from the point of view of some very clear-cut "inside instrumentalists." With this interest in mind, we turn to the edited volume *Learning in Development Co-operation: An Introduction* (Carlsson and Wohlgemuth 2000), which provides an interesting reference point for research that deliberately seeks to use academic insights for the improvement of development aid operations. That volume, directed at an action-oriented audience, seeks to create a "multi-actor perspective" on development in order to "maximize policy relevance" (6). Yet the editors, one of whom once worked for the Swedish International Development Cooperation Agency and the African Capacity Building Foundation, are by no means sanguine about the challenges of facilitating such policy-relevant learning processes. Nevertheless, *Learning in Development Co-operation* is intended both to point to problems and to present practical solutions.

From their "multi-actor perspective," the contributors to *Learning in Development Co-operation* take the stance that not only bilateral and multilateral aid organizations but also a variety of actors at the receiving end of aid projects should be viewed as stakeholders in development aid. When contributors to the volume link the problem of organizational learning to the issue of aid effectiveness, prefiguring later mainstream policy agendas such as the Paris Declaration and the Accra Agenda for Action, learning from the varied perspectives of aid recipient governments, organizations, and local people about the qualities and characteristics of the aid process is significant. One reason this is interesting to us is that learning from such absent others is also an important problem for technologists involved in building the infrastructures of accountability and transparency that we study.

In *Learning in Development Co-operation*, learning processes are seen as usually problematic. Ideally, learning entails that substantial information is "created on both sides" (Carlsson and Wohlgemuth 2000, 8) and is used by all involved actors; in reality, however, the authors estimate, this happens too rarely. For example, Elliott Berg, a professor of economics and an experienced development consultant, circles around the issue of "evaluation shortcomings" and points to problems relating to the "logical framework concept" and to "performance indicators" (31–33).[4] Other contributors zoom in on evaluations and what fails to be learned from them, an issue that Anna Wieslander connects with the fact that "bureaucrats notoriously lack time" (2000, 258). Gus Edgren, Sweden's ambassador to Vietnam at the time of writing, draws on the management guru Peter Senge's concept of learning blockage in identifying a series of "obstacles" that relate to internal organizational issues, including team learning, staff composition, incentive systems, and leadership. Willi Wapenhans, formerly a senior vice president at the World Bank, expands the analytical concern by pointing to a "lack of coherence in development policy" (2000, 229), noting that "the insistent call for accountability and greater effort in measuring developmental progress suggests dissatisfaction with results thus far achieved" (229). This certainly is a prescient insight, in view of the subsequent increase in interest in both accountability and aid effectiveness. But Wapenhans doesn't dwell long on the point or connect it with broader cultural or political analysis.

Overall, *Learning in Development Co-operation* focuses on particular kinds of problems that we can broadly designate as organizational and operational. Nevertheless, these analyses also overlap in important ways with diagnoses offered by critical anthropology. For example, they point to a variety of "political constraints" that impinge on the aid relationships, and they are explicit about the unequal nature of such relationships. In addition, they see problems as simultaneously internal to the aid organizations themselves and built into the relationships these organizations have with diverse organizations on the receiving side.

Noticing the resonance of these suggestions with critical analyses of development is, of course, different from claiming their identity. If our reading of *Learning in Development Co-operation* suggests that development practitioners are not as naive as they are assumed to be by critical development anthropology, it is also obvious that their applied concerns confine the analyses offered by practitioners in ways that leave many crucial questions about development aid and its consequences unanswered, or even unasked. They are, indeed, working from within the development doctrine.

For example, Elliott Berg wonders why organizations seem not to learn from their mistakes and why they usually respond inadequately to failures. His answer to this question takes the form of "factors" relating to "obstructing elements" in the environments of aided countries, "shortcomings" in formal evaluations, and the characteristics of aid organizations themselves (2000, 29). Yet while these elements may contribute in important ways to "failure," we can still pause to note the narrow vision that underpins this definition of the problem space. Indeed, the "obstructing elements" in the environment of an aided country is likely to be deeply intertwined with national, social, economical, and political systems, if they are not actually coeval with such systems in their entirety (Rottenburg 2009). In that case, these "elements" would certainly pose significant obstacles to learning and improvement. But such obstacles would go considerably beyond the technological and organizational concerns identified by Berg.

Following Berg's observation, we may further ask "Precisely who views what as obstructing what?" The question is whether actors in the receiving countries would agree with the specification of obstacles, or whether they would, in fact, point to an altogether different set of obstacles—problems that may themselves be either invisible to or perceived variably by different development actors. Such questions, however, are outside the scope of *Learning in Development Co-operation.* But neither are they adequately addressed by a critical ethnography that takes as its task unveiling the ideologies and errors of aid organizations.

Accordingly, we are in need of a third position that will allow us to differentiate not only aid development as analytical topic (Yarrow and Venkatesan 2012) but also the complex relationships between development partners at multiple local and global levels. Indeed, it will have to be able to deal with the troublesome problem that both organizational and analytical scales are subject to change due to the ontological displacements of infrastructure. With this in mind, we proceed to carve out a position which recognizes that neither infrastructures nor partnerships add up to wholes, and that new gaps in knowledge are invariably created in the very process of filling existing gaps. From this acritical position, we see infrastructures and partnership as effects of divergent practical ontologies.

Co-Evolving Infrastructure and Sociality

In a 2004 article on development interventions in India titled "Is good policy unimplementable?" David Mosse asks rhetorically whether anthropologists have "really learnt all that we can" from development projects (2004, 640). Certainly no one could accuse Mosse of a lack of reflexivity. Yet note how his question almost literally replicates the one posed by the Carlsson and Wohlgemuth in their explicitly *applied* discussion of learning and development. Mosse, too, offers the same general answer: No, anthropologists, like development workers, have not learned all they could.

Directing his analytical interest to interrelations of policies, practices, and infrastructures, Mosse argues that the primary function of policy is not and cannot be to "orientate practice," because it is, rather, to mobilize and maintain support for, and thereby the legitimacy of, aid projects. Development interventions, in his view, are therefore never driven by policy; instead they are driven by the "need to maintain relationships" (651) in complex networks that comprise not only people but also systems of representations.

The anthropologist Julia Elyachar (2010) offers a complementary entry point for considering the relationship between infrastructure and development work. Studying empowerment among women's organizations in Cairo, she connects infrastructure to the theme of "phatic labor" in an analysis of the "social infrastructure of communicative channels" (452). Elyachar sees such channels as "outcomes of practices of sociality" that can be engaged as distinct objects of inquiry (455). Instead of focusing on how people are "brought together in networks" (a typical point of interest for STS and information studies), which would imply a focus on the network and the way it affects sociality, Elyachar points to the anthropological importance of analyzing how communicative channels are themselves fairly stable results of social practices.

These arguments link back to our practical ontological engagement with infrastructure. Mosse highlights the power of infrastructures once they have reached a certain level of stability (once their "base" has been "installed"). For example, infrastructures may affect how development organizations work, may redefine relations between organizations, policy makers, and other collaborators, or may turn villagers into partners. Juxtaposing Mosse and Elyachar, we end up with a simultaneous focus on the sociality of infrastructure and on the ways in which such sociality is itself infrastructured. (See chapters 5 and 6.) This is a situation, then, in which sociality and infrastructure co-evolve.

In his elucidating and entertaining study, Richard Rottenburg discusses a related form of socio-technical co-evolution in a context that bears directly on the aid transparency agenda and the infrastructural efforts that we study in subsequent chapters of the present book. Rottenburg suggests that "all the different organizational variants of the development enterprise are concerned with a simple but extremely tricky problem." He continues: "If I lend money to someone who then uses it legally to earn more money and I in turn am paid interest, then this interest is sufficient proof that it was a good investment. If, however, I lend money belonging to a third party to someone who proves unable to make a profit with it, then I need a different kind of proof to demonstrate that this was nonetheless a good investment." (2009, xvii) Rottenburg shows in great detail that creating such visible proofs is indeed extremely tricky for all involved parties. However, development workers as well as critics recognize this trickiness only too well. Indeed, this is one of the main reasons why the idea of creating infrastructures to facilitate a smooth and transparent flow of information is so appealing. (See chapters 5 and 6.) Even so, most visionaries and builders of such infrastructures—and most critics—pay little attention to the co-evolution of formal systems of representation, infrastructures of accountability, social and organizational practices, and new forms of aid "users" and aid politics.

Rottenburg proposes to view development projects as an "archetypical case of *distributed agency,* in which several actors, dispersed over different locations and social worlds, possessing different webs of belief and conceptual schemas, have to collaborate in order to achieve something" (xxvi). He argues that too little attention is paid to "interstitial spaces" of aid infrastructures. The interstices of infrastructure are relational spaces that formality doesn't reach, and in which interactions are therefore not structured by demands for transparency. They are the gaps that are left unaccounted for by standardized systems of representation. Whereas Mosse shows how new or conflicting demands relating to changes in policy networks can radically transform perceptions of specific local projects (changes coming from the outside in, as it were), Rottenburg argues that "the practice of development is . . . primarily a matter of selecting one of the existing globally circulating and highly esteemed models for development and adapting it to a local context" (xxvi)—that is, picking the right kind of outside to bring in.

But perhaps this choice between being changed by an outside and choosing one is too stark. This is precisely the import of our argument: that the outsides and the insides of aid projects transform through processes of co-evolution. And, that to a significant extent, they do so through infrastructure.

Friction and Noise

In recent years, the notion of cultivation has been proposed both in infrastructure design (Aanestad 2002; Hanseth, n.d.) and in development anthropology (Mosse 2005b) as a way of understanding collaborative challenges and possibilities. The notion of co-evolution outlined above and the concept of cultivation seem to fit well together, insofar as both of them point to the gradual creation and nurturing of a common space in which issues can be addressed and solutions agreed upon. Co-evolution and cultivation differ in one important way, however: the latter implies a sense of harmony and consensus that is lacking in the former.

If development projects co-evolve among heterogeneous actors, they do so, as the philosopher of science Isabelle Stengers says, without converging. Different "stakeholders" continue to "go on defining in their own manner what matters for them" (Stengers 2011a, 60). From the point of co-evolution, therefore, the central question has little to do with how to ensure consensus; it has to do with the continued negotiation of divergent and possibly incommensurable interests.

The anthropologist Paul Kockelman offers a set of concepts that relate to divergent co-evolution in infrastructure. In an article titled "Enemies, parasites, and noise: How to take up residence in a system without becoming a term in it," he defines infrastructure as "a condition of possibility for the relative comparability of value judgments across actors within a community who are using, exchanging, and representing things during semiotic events that are relatively displaced from each other in space, time, and person" (2010, 419). Kockelman emphasizes that infrastructure is in some sense about facilitating comparability. In chapter 5 we will see this played out in an infrastructure that links a Danish environmental NGO with a project in Vietnam. As Kockelman argues, it is extraordinarily difficult to create comparability in just the contexts in which it is most needed—that is, in settings in which actors are at once heterogeneous, dispersed, and in disagreement about fundamentals. Thus, Kockelman considers the tension between incommensurable value systems and standards. Such tensions are often rendered invisible by discourses on transparent partnerships that speak as if we are all "in this together." They also fade from view if one approaches infrastructure and development as projects of shared cultivation. Kockelman argues that infrastructures involve processes that create animosity, parasitism, and noise. Taking us to the "dark side" of infrastructure development, his argument suggests that the co-evolution of infrastructure and sociality invariably also entails the creation of new forms of disconnection, disagreement, or even hostility.

Contrary to what might be argued by "instrumentalists," these are not unfortunate consequences of badly planned development. They can be avoided neither by more avid adoption of participatory methods nor by designing more truly transparent technologies. Rather, infrastructural noise and disconnection must be seen as more or less endemic, unavoidable outcomes of the very effort to create partnering, fair, transparent networks. As the STS scholars Marc Berg and Stefan Timmermans argued (inspired, like Kockelman, by Michel Serres), the construction of an order—such as a standard or an infrastructure—always implies the simultaneous construction of its very own disorder. (See Berg and Timmermans 2000.) Thus, new forms of transparency co-construct new versions of opacity, and new ways of doing accountability also create zones of obscurity into which formal accountability doesn't reach. The submerged effects of development crop up in these interstitial zones. These effects proliferate alongside the introduction of new standards, formats, and technologies, all of which become progressively infiltrated with heterogeneous practices of globally distributed sociality.

Kockelman's metaphors of animosity, parasitism, and noise trouble the image of harmony and order that lurk behind aspirations to cultivate social and technological practice. Yet even more interesting for our purposes is Anna Tsing's (2005) notion of *friction*. Friction points to the co-existence, in any development project, of multiple and only partially connected dreams, schemes, plans, and aspirations. It indexes the noise and heat generated in heterogeneous collaborative settings, but doesn't view them as exclusively negative. Rather, friction requires us to think about the multivalent, ambiguous effects of aid partnerships and infrastructures. It helps us think about both "yearnings and nightmares" (Tsing 2005, 1) in infrastructural encounters.

Attentiveness to infrastructural friction and to the noise of partnership helps us stay clear of the infrastructural dream language that turns development collaboration into a purely communicative or technical issue without demanding an analytical *salto mortale*, whereby we obtain a purely critical perspective. Aid information infrastructures may enable new communicative practices, to be sure, but only by also reconfiguring modes of sociality, and only by creating "new gaps even as it grows through the frictions of encounter" (2005, 13).

The historian of ideas Georges Canguilhem reminds us that in the nineteenth century the words *progress, perfectibility,* and *development* were seen to be "interchangeable." However, "the basic meaning of *development* is . . . the dominant term: the successive manifestations of the potentialities contained within the germs of an organism" (1998, 317; cf. Cowen and Shenton 1996). STS and anthropological studies of infrastructure and aid development have demonstrated that this understanding of development as the organic unfolding of potentials through a linear, orderly, and perhaps cultivated process has little to do with any "actual development" of either technological or organizational systems.

If, as Canguilhem argued, history in general has given the lie to "the optimistic claim that its course was progressively rational" (1998, 316), both the STS literature on infrastructure and the anthropological literature have made the limits of viewing development in terms of "progress" and "perfectibility" abundantly clear. The following chapters on aid information infrastructures in friction provide ample illustration.

Practical Ontology and the Problematic Landscape of Aid Infrastructures

In this chapter we have positioned our approach in relation to science and technology studies, information systems research, and the anthropology of development. In the first half, we described how infrastructures have increasingly come to be understood as socio-technical, rather than merely technical. We also elucidated how information sharing ties in with particular understandings of infrastructures as social devices for collaboration and knowledge making and as techniques that reshape the social. In the second half, we engaged with the history and anthropology of development in order to place the current interest of aid development organizations in information, accountability, and transparency in a historical and an analytical context.

Having introduced important analytical and theoretical heritage enables us to locate our work on aid information infrastructures as part of a problematic landscape, which can be labeled as the exploration and elucidation of practical ontologies. Conceptually and methodologically, the study of practical ontologies is inspired by, and closely related to, actor-network theory and affinal approaches in STS (Latour 1987, 1988, 1999; Mol 2002; Pickering 1995; Verran 2001) and the anthropology of Marilyn Strathern (1991), Annelise Riles (2000), Bill Maurer (2005), Mei Zhan (2009), and Eduardo Viveiros de Castro (2004).

As is well known, actor-network theory takes the study of relations as the basic analytical unit. Doing so, ANT dispenses with any *a priori* delimitation of what can count as a relation. For ANT, relations "emerge" and "grow" in networks, which gradually may "gain in reality." Thus, the scales of a network are never given but always contingent. Further, networks themselves become actors. This is therefore not an issue of perspectives "on" reality. For ANT, what is growing, emerging, and transforming *is* reality. The level of analysis is thus ontological rather than epistemological.

Learning both from actor-network theory and from social and cultural anthropology, practical ontology is at once a description of the work done in the world by heterogeneous actors (such as the many people, technologies, and organizations engaged in monitoring) and a description of the work done by researchers aiming to elicit that work. The practices described in this book are brought into being through practical ontology; they come together in what we call the *monitoring movement*. The monitoring movement "clots" through concrete empirical "links" and through the comparative, conceptual work that allows us to see how relations cut across disjoined networks. Rather than choosing between emerging networks and stable forms, practical ontology enables us to keep difference and consistency, and the empirical and the conceptual, alive together.

2 Recursions: Partnerships, Infrastructure, and Ethnography

Partnership as Form

In the previous chapter, we made note of the fact that the notion of partnership emerged at a particular historical moment, crystallizing in the Pearson report. (See Power 2003, 132.) The Pearson report conceptualized aid in terms of partnership. Although at the time "partnership" referred mostly to governments, the report also had profound implications for how people began to think about the "goods" of aid.

Today partnership is a salient feature of aid cooperation, if it isn't the most prominent and dominant form through which development programs and practices unfold. Partnership in its current guise, however, is far different from the nation-to-nation relations envisioned by Pearson and his co-authors. At present, it has to do with "putting aid-recipient countries in the driving seat" and ensuring "local ownership" (Mosse 2005a, 10; see also Gould 2005, 65).

The fact that the present discourse of partnership paints a rather harmonious view of how donors and recipients relate, one that has relatively little to say about economical or institutional power relations and asymmetries, obviously doesn't mean that such relations no longer have a role in development aid. Critical development researchers (e.g., Cooke and Kothari 2001) have analyzed partnership as a particularly insidious rhetoric precisely because it seems to hide power under an ideological smokescreen of equality, commonality, and shared win-win situations. However, this bleak view misses much of what is actually taking place under the form of partnership.

In any case, current forms of partnership involve a far more diffuse set of actors than those engaged in macro-politics. Indeed, partnership has become the ideal for cooperation by governments as well as for grassroots efforts. It provides a way of describing practically any set of relationships in development work. For example, farmers are expected to partner with farmers, but also with non-governmental organizations' field staff and researchers, who are expected to partner with each other. NGOs are expected to partner with each other, and also with businesses, cooperatives, or governmental institutions (Lewis 1998, 108).

In our cases, developers of aid information tools and platforms likewise partner with a wide variety of actors who, meanwhile, constantly look for yet more potential partners. Even Audit Offices engage in participatory exercises that implicate them in processes of organizational learning with the ministries and agencies they audit. For social scientific researchers, such as ourselves, too, partnership becomes a condition rather than merely an object of study.

As these varied examples indicate, there is at present no universal model for partnership, and there probably never has been one. But we argue that even though partnerships are very different, consisting of different actors, with different goals, who engage in different forms of collaboration for different purposes, partnership is itself invested in particular forms of relation making (Thévenot 1984), much as the anthropologist Annelise Riles (2000) suggested that a particular aesthetic form undergirds discourses and practices of networking. In her study of international NGO networks, Riles noted that "the idea of the Network, as the term is used here and by the subjects of this study as a form that supersedes analysis and reality, might also be imagined to borrow from the reflexive turn in the social sciences—from the notion that there is no longer such a thing as dependent and independent variables, that causes and effects are all mutually constituted in an endless feedback loop" (2005, 263). Riles realized that the discourse and analytics of networks and networking were as important to the subjects of her study as they were in anthropological theorizing.

In our case, much as it was inconceivable for Riles' activists to live outside networks, it seemed impossible to refrain from thinking through and engaging in partnerships. Indeed, this entails doubt as to whether it is possible to study these activities and practices without becoming a partner. The upshot of this is that our social scientific knowledge claims in this context are encompassed by the partnerships we aimed to study. This raises questions about the very form knowledge can take in this context. In short, and recursively, partnership has consequences for the making of knowledge about partnerships. While partnerships, like networks, promise to provide development actors with simply "a technical device for doing what one is already doing, only in a more efficient, principled and sophisticated way" (ibid.), they thus do much more. Among other things, as Riles noted for networks, partnership can "manufacture . . . desire through mundane 'technicalities'"; it has "the activating power of unnoticed forms" (265).

This activating power is at play both in fairly obvious settings (for example, aid projects and programs in which NGOs and their counterparts in the ministries, organizations, and villages of developing countries engage) and in settings in which it is less clear who the partners are and why they would partner with one another in the first place. Why, for example, would political scientists interested in aid development come to partner with philanthropists and NGOs? As we will show in chapters 3 and 4, they do so both to advertise and disseminate their own database designs and to find out how to make it meaningful to a different set of not-yet-actualized partners: the enigmatic aid recipients on the ground. Partnership also positions these researchers in relation to a host of other actors and turns them into scientists with money, strategic goals, and connections to powerful development organizations and donors. Likewise, partnership is changing dominant audit paradigms, even in such bastions of "mechanical objectivity" (Porter 1995, 4) as the Danish National Audit Office. As we have noted, partnership's demands for specific arrangements, promises, and terms of reference have conditioned our own knowledge making.

Encompassing Partnership

In *Expectations of Modernity*, James Ferguson suggested that the "metanarrative of urbanization and modernization," which most anthropologists used to make sense of their ethnographic data on urbanization in African contexts, was itself part of the ethnographic data that had to be made sense of (1999, 16). In a context of encompassing partnership, our observation is similar. Partnership is not only an empirically observable phenomenon. It is also a set of analytics, a means of engaging current development practices. Indeed, partnership is an analytical resource already embedded in the field we study, as exemplified by the volume *Learning in Development Co-operation*, which we discussed in chapter 1.

The fact that partnership is simultaneously "external" and "internal" raises questions about the status that research produced "within" the field of development might have in relation to our own. For example, we might wonder whether analyses offered in volumes like *Learning in Development Co-operation* are part of our empirical data or, rather, whether such volumes offer theories and analytical tools we might ourselves make use of. The point is that in a context of recursive partnerships this question has no definite answer. Indeed partnership not only encompasses the relation between "theory" and "data" but also extends to our practice of research. This situation has implications of more general valence for current social scientific efforts to distinguish the empirical from the conceptual. (See Gad and Ribes forthcoming and Jensen forthcoming.) It also has implications for naive understandings of how scientific research might benefit "industry" or "society."

In each of the settings we have mentioned, partnership exhibits the loop-like quality that Riles remarked upon in the context of networks: "causes and effects are all mutually constituted in an endless feedback loop." We find such loops again and again, embedded in our empirical material, influencing how we had to think methodologically, and shaping our analyses. This is why in what follows we talk of recursive partnerships, recursive infrastructure, and recursive ethnography in the same breath.

To get a sense of the extent to which the form of partnership permeates divergent settings, contexts, and knowledge-making practices, we continue to offer a series of illustrations. Our choice of illustrations is meant to invoke the range of empirical settings in which partnership permeates monitoring practices. More important, it is meant to indicate the breadth of analytical themes that are conjured once one looks in more detail at how and where partnerships are enacted. Thus we point to four versions of partnership: as audit loops, as conflicting demands, as multi-directional gate keeping, and as divergent knowledge making. Each of these versions also illustrates different ways in partnership is interrelated with the design and the use of infrastructure.

The vignettes below highlight how partnership emerges from recursive interactions that cannot be placed within a predefined empirical or conceptual domain. These versions of partnership have thus also been chosen to emphasize that partnerships and infrastructures recur across empirical and analytical "levels" of development and thus do not belong to any particular scale at all. (See Jensen 2007a.)

Partnership as Auditing

The National Audit Office is an independent institution under the Danish Parliament. Its specific function is to audit the budgets and the expenditures of ministries and state organizations. This means both that the institution verifies financial accounts and that it conducts comprehensive performance reviews in order to examine whether state funds are being used for the purposes decided on by Parliament and by the audited organizations. Thus, the explicit institutional mandate is to control and evaluate organizations.

One might reasonably wonder what this has to do with either aid development partnership or infrastructure. But among the National Audit Office's tasks is to conduct performance reviews of the Ministry of Foreign Affairs and of Danida. It specifies the documentation that ministerial departments must provide, and it audits the performance of organizations on the basis of that information. In performance audits of aid-related activities, this can involve visiting Danish embassies and monitoring selected projects supported by Danida in developing countries. In this particular context, auditors, having adopted the lingo of the consultants they monitor, occasionally refer to their audits as fieldwork.

The Audit Office appears to be an exemplary external agent. Indeed, it is crucial for the institution to be perceived as a non-partner, because its institutional mandate is to provide objective knowledge on accounting matters to Parliament and to politicians. Nevertheless, more collaborative forms of interaction between auditors and those they audit have made headway into its organizational practices and ways of thinking. It isn't simply that notions such as collaboration and mutual learning are used colloquially and informally; rather, the language of partnership is adopted even in official reports. For example, the annual report for 2008 highlighted a new holistic concept for auditing that was developed to create value for businesses by offering an overview over their strengths and weaknesses. While the control function remains central, a partnership dimension is introduced through the emphasis on the sharing of knowledge about best practices. In chapter 6 we argue that this situation is symptomatic of a gradual transformation of the audit paradigm. We call this transformation *audit implosion*.

More generally, international trends in development aid promote the devolution of managerial responsibilities from donors, such as Danida, to governments in developing countries. As we noted in chapter 1, this implies a gradual change from project-specific funding to budget support, with large sums of money transferred directly to ministries in developing countries. One of the challenges faced by the National Audit Office is that such transfers render this money largely invisible to its controlling gaze.

Though one might imagine that the parliamentary mandate of the Audit Office—to ensure the proper use of Danish taxpayers' money—is put under pressure when funds are allocated through budget support that cannot be audited directly, that isn't how the auditors we interviewed saw it. As far as they were concerned, if development experts and politicians encourage budget support it is not their task to judge the merits of this strategy. Instead, the office must ensure that monitoring systems for measuring effects and stimulating self-monitoring and learning inside these organizations are in place. Developing such systems, however, engages the office in a counseling role, with auditors providing advice. It may, for example, advise the Ministry of Foreign Affairs on how to monitor and evaluate in difficult and uncontrollable settings, such as foreign bureaucracies.

However, this blurring of the roles of counselor and controller comes with certain built-in dilemmas. What happens, for example, if the Audit Office suggests improvements in the monitoring systems used by the donors, not as a result of an audit that leads to a formal reprimand and a call for improvement, but rather as a way to help nurture organizational learning? This situation is potentially troublesome, since subsequent audits will then, to an extent, monitor the effectiveness of solutions provided by the office itself. The office will then have entered what we refer to as an *audit loop*. In such cases, aside from questions concerning objectivity and legitimacy, the office may be seen as responsible for evolving problems in the organization.

Partnership in the National Audit Office is exemplified through the stepwise and tentative adoption of a counseling role that complements the traditional role of controller. As we will discuss in more detail below, this role cannot be disentangled from the infrastructures of information sharing on which the office relies in making knowledge. Partnership as audit loops, then, points to the ways in which auditors become recursively implicated with what they monitor even in a context in which objectivity remains an unquestioned value.

Partnership as Conflicting Demands

If the National Audit Office finds itself engaged in partnership-like modes of engagement somewhat in spite of itself, the same cannot be said of the enactment of development collaboration between donors and aid recipients. Partnership as defined in the Paris Declaration explicitly targets these actors. Neither the Danish embassy in Hanoi (which we visited) nor any of the Vietnamese ministries with which it collaborates doubts that it is involved in partnerships. But the forms of collaboration this entails are still emerging and uncertain, subject to negotiation and revision. Our next example exhibits some fundamental uncertainties concerning the ontology of partnership in these settings. What does partnership entail for the partners? What demands can they make on one another? What consequences do such demands have? Consideration of a report and interviews with a consultant from Australian Aid (AusAid), with three government officials at the Danish embassy, and with a British consultant working for Danida in Hanoi allow us to address these questions.[1]

In April 2008, government representatives from Laos, Cambodia, and Vietnam met in Bangkok to conduct what were called the South East Asia Consultations. The consultations took the form of an informal meeting that led to a report submitted as a contribution to discussions on mutual accountability between donors and their counterparts for an upcoming meeting in Accra, Ghana—the Third High Level Forum on Aid Effectiveness. The report was based on the assumption that a "genuinely partner country-led contribution on mutual accountability"[2] would be important. The initiative was, in part, a way of addressing the fact that mutual accountability was seen to have unclear implications both in theory and practice—that, in the words of an AusAid consultant, it was one of the "least understood principles of the Paris Declaration." Here we indicate how discussions of the implications of mutual accountability enabled the articulation of conflicting demands among partners.

In advance of the High Level Forum, lots of preparations were taking place in organizations and networks around the world. Yet there was a serious concern on the donors' part that not enough input was being generated by the actors actually receiving the aid. They, not the donors, were supposed to be setting the agenda. Since no such contributions appeared to be forthcoming, however, the initiative to hold the South East Asian Consultations was taken not by the countries themselves but by a representative of the United Nations Development Programme. In turn, the suggestion to have a consultation meeting was taken up quite differently by the invited governments. Whereas Vietnam saw an opportunity to set the agenda, Cambodia showed little interest—perhaps because Cambodian officials had learned that they were to be parties in the consultation by reading the minutes of the meeting at which the decision had been made in their absence.

In the event, the Southeast Asian countries made the most of the opportunity to provide input. They proposed to "develop a set of recommended priority actions to remove obstacles" to the achievement of mutual accountability, arguing that the process in which they had participated provided a model for "south-south capacity development" (that is, learning and collaboration among developing nations). And they agreed on a common platform of recommendations.

It is particularly instructive to note how "obstacles to behavior change" in relation to establishing mutual accountability came to be understood by Vietnamese participants. In their interpretation, almost all problems were caused by the donors. Thus the report noted that "in spite of the government leadership on MA [mutual accountability], it appears that there has been very little guidance . . . from the donor headquarters to their local offices in terms of MA." It also suggested that some measures of mutual accountability were residual conditionalities belonging to the days of structural adjustment rather than to the paradigm of partnership. The Vietnamese report seemed in broad agreement with the historian Fernand Braudel's rather cynical "theory of the donor" (1976, 826), according to which "he who gives, dominates." This is precisely the situation they wanted to change.

The Vietnamese report pointed out that "no donors inform about their strength and the way of delivering aid (project and non-project types)." This, the report stated, made it hard for the Vietnamese government to develop a coherent and rational framework for handling the inflow of money. Thus it was emphasized that, although a development assistance database that purports to be comprehensive exists, donors still provide inconsistent data at irregular intervals. Finally, the challenge of aligning the policies and strategies of donors and recipients was also placed squarely in the hands of the donors, who should not wait "until the partner countries reach international standard but align where the capacity exist."

What is striking in this example is how, contrary to the presumptions and ideals of the Paris Declaration, the Vietnamese delegation didn't view partnership as a benign and consensual format for accountable, transparent collaboration.[3] Clearly, an aesthetic of partnership was present and enacted in the meetings held and documents that were circulated in advance of the Accra High Level Forum on Aid Effectiveness. But aside from setting up a particular form of collaboration among donors and aid recipients, partnership became a platform for specifying points of contention and for formulating new demands. These demands related to the collaborative formats used by donors, their internal and external information and communication systems, and their budgeting and evaluation systems. Meanwhile, newly implemented technologies, including development assistance databases, enabled the Vietnamese to make the case that donors had failed to live up to their obligations.

Using Deborah Heath's word, we can characterize the Vietnamese response to partnership as "revealing" (1998a, 71). What it reveals is how partnership, when put into global circulation, gives rise to processes of reinterpretation, appropriation, and exchange that are "grounded in local knowledge" and provide "grounds for transformative translocal practice." For the Vietnamese delegation, in particular, partnership became a platform for highlighting tensions and for formulating demands to be made of "translocal" donors. Partnership became a platform for emphasizing friction, for suggesting that not all parties were in agreement. Recursively, partnership enabled the Vietnamese to open a discussion about what partnership itself meant and about how it obliged partners to act.

Partnership as Multi-Directional Gatekeeping

Our third example illustrates how we became caught up in partnership as part of our research. That this wasn't a voluntary matter suggests something about partnership as a particular mode of collaboration and knowledge making.

In October 2008 we went on a field visit to Hanoi. We had arranged to meet with employees at the Danish embassy for the purpose of interviewing them about the bureaucratic and infrastructural aspects of Danida's work in Vietnam. We hoped to make contact with people working on Danida projects in Vietnamese ministries. We also hoped to visit sites where these projects were implemented. However, it turned out to be rather difficult to make such arrangements and the issue of partnership was significant in that regard.

Employees at the embassy repeatedly emphasized that getting access to interview persons and visit field sites would be very difficult. They explained that they didn't want to put pressure on their partners in the ministries by introducing researchers whose studies were not closely aligned with their immediate goals and objectives. In the past this might not have been a problem, they indicated, but under the form of partnership it was. One reason was that a major side effect of partnering was that government officials and embassy employees were increasingly busy implementing monitoring and evaluation systems and arranging meetings with multiple stakeholders. More than once we were told that it would be far easier to do our research in Africa. The message seemed to be that in Africa donors could propose or perhaps even dictate what partners should or shouldn't do in a manner that was inconceivable in Vietnam.

Had partnership provided an effective platform for the display of Vietnamese recalcitrance to demands that would place the country in the role of passive recipient? Was partnership simply a delicate way for donors to get around the often-remarked-upon Vietnamese "independence," or was it a way of dealing with the fact that, as a Danida employee said, "donors and their counterparts are completely different animals"? Paradoxically, from the point of view of the embassy's officials, the independence of the Vietnamese *as partners* turned them into simultaneously better *and* worse collaborators than the Tanzanian partners with whom some of the officials had worked previously. The overall suggestion was that if we were to have a chance of getting anywhere with our research we too would need local partners. The embassy, unfortunately, could not be of assistance.

The importance of finding local partners was also emphasized in later conversations. Vietnamese and Danish economists, for example, worked together on macroeconomic issues, and Danish visitors intermittently set up shop in Hanoi to teach economy courses to Vietnamese students. Anthropologists who had worked in Vietnam had been able to do so effectively by hiring local university students as field workers. A highly experienced Danish biologist working for the Danish Development Research Council told us that not engaging in partnership would be unethical. In a way, of course, we could only agree. Thus we found ourselves increasingly encompassed by the context of partnership that we meant to study, as it were, externally. But this change in approach was neither of our volition nor due to an analytical or political predilection. It was a practical circumstance relating to an ethnographic "absence of control" that, as James Ferguson noted, "may not be so pleasurable" (1999, 18).

If we wanted to study the infrastructures of accountability that tied together development partners, then we too would have to become partners, and this would imply "alignment" of our research goals and objectives. At the time we were neither able nor willing to re-align our deliberately open-ended investigation. For this reason, we didn't manage to establish a partnership relation with Danida or the embassy. In that specific sense our effort failed. Yet our encounters with people at the embassy, our occasional conversations with aid consultants in Hanoi, and our interviews with Vietnamese aid practitioners and with Danish scholars in Vietnam *were* important pieces of our fieldwork. These conversations offered valuable insight into partnership in their own right. But they also set us off on the path that has taken us to where we are now. This is a position from which we are able to characterize present-day development aid in terms of practical ontologies through which infrastructures of accountability, partnership, and knowledge making intersect and co-evolve.

This episode, then, illustrates how partnership implies gatekeeping. Partnership both prevented and enabled our field access. As researchers we had to be seen as trustworthy and relevant partners to be able to enter the gates of ministries and organizations, but in the situations just described the terms of acceptability were not ours. We do not state this as a complaint, though it wasn't always pleasurable. To be sure, being accepted as a partner wasn't always pleasurable either; for one thing, it made conducting interviews with aid recipients and other local "partners" very difficult.

However, the central point is ethnographic, because the experience explicates important dimensions of what partnership entails. On the one hand, it reiterates the point (often made by critics of development) that partnerships are never quite equal, and that the term hides conflicting demands underneath consensus rhetoric. More centrally, however, it suggests that partnership, rather than creating a transparent communicative space, creates multiple gates that swing in different directions and are controlled by different actors, depending on time and place. But this also illuminates another sense of recursive partnerships. It highlights that partnership creates situations in which *everyone* is dependent on the gatekeeping activities of others. Thus, it points to situations in which decision power is distributed and channeled through a labyrinth with multiple entrances, dead ends, and exits (Jensen 2010b).

Now, if one must be a partner in order to study partnerships, and if partnerships create a maze with many gates and exits, it follows that the researcher cannot be alone in determining what matters. Had we managed to established partnership with Danida or with its Vietnamese partners, we would certainly have entered the labyrinth. But though we didn't establish this particular partnership, we still had to enter the maze. Only it would have to be by a different route.

This is the sense in which knowledge about partnerships can't be gained from an outside position. One must enter the labyrinth in order to gain any knowledge, but upon entering it one is obliged to enter a game of negotiation that alters the pathways on which one can travel. Whatever the entry point, partnership knowledge is constructed as part of the emerging process of enacting partnership itself. Thus, the following chapters show partnership as entailing a form of relationality in which partners' agendas and capacities, including our own, are mutually made and remade.

Partnership as Divergent Knowledge Making

Partnership also involves processes of divergent knowledge making. Partnership and divergence are not mutually opposed categories. Recall Isabelle Stengers' argument that co-evolution entails partnership precisely insofar as the involved parties pursue divergent interests through one another. Gilles Deleuze and Félix Guattari's famous example of the mutual becoming of the wasp and the orchid can be used to illustrate a divergent partnership that works, not by linkage, by mixture, or by agreement, but rather through "shared proximity" (1987, 293–294). Shared proximity, rather than consensual collaboration, allows the wasp to gain food from the orchid, while the orchid spreads pollen via the wasp.

In several of our cases, too, partnership involved sharing proximity in ways that allowed for divergence. Here we point to some examples that focus on how various kinds of proximity facilitated divergent knowledge making. Indeed, divergent knowledge making may entail a redistribution or a lateralization (Maurer 2005) of epistemic authority that is particularly interesting for ethnographic method and theory.

During a 2010 conference on aid data, transparency, and corruption (a conference that we attended in the course of our fieldwork), we found ourselves in a pub in Oxford with a group of political science researchers who were involved with a new database called AidData. The public release of AidData was to take place at the conference. One principal investigator was fairly stunned to realize that, for us, he and his team were objects of study. Subsequently reflecting on the meeting, he wrote a blog entry titled "I am the data or: How I went from gathering data to being a data point."[4] He used the blog entry to think aloud about different scientific communities with differing knowledge paradigms:

> There are normally not a lot of anthropologists hanging out at the conferences I attend. It turns out that anthropologists have long abandoned the "fly on the wall," they engage their subjects on their own terrain, watch them interact with other people and with "objects" in their environment, take lots of notes (but not in bars), and even tell their subjects about the nature of their research.

As his comments make clear, the ethnographer is enrolled as his informant in a manner that facilitates a certain kind of reflexivity. Intrigued by the exoticism of ethnography's participatory methods, the blogger brings to life the encounter between the ethnographer and the informant. The ethnographer becomes a proxy for our informants' conceptualization of ways of doing science and, indeed, of the very nature of data. A lively characterization of his assumptions and newly discovered realities about the knowledge-making practices of anthropologists figures prominently in his description. "It was both weird and fun to be a data point, rather than the person studying data points," he writes, adding that the method of "beer talk" may cause him to rethink his approach to surveys.

The aforementioned blog offers a rather entertaining example of how ethnographers are not only "snowballing" research sites but also "being snowballed" into new ones. In the setting described, partnership was informal or even metaphorical. Our next illustration, however, deals with a far more overt partnership relation that nevertheless exhibits similar features of recursive knowledge making for divergent ends.

One evening, during a review visit to an ecotourism project being conducted by NatureAid, one of us (Brit) was at a national park in Vietnam. In the course of an entertainment program, some villagers put on a "cultural performance." The photograph reproduced here as figure 2.1 shows Brit joining the dance. This "cultural performance" was documented by our main informant, who had been trained as an anthropologist before taking a position as a project manager. From an anthropological point of view, the photo displays something like an archetypical encounter between a "local" member of a culture (a villager, for example) and a newcomer—typically an ethnographer, who, as a scientist represents a "culture of no culture" (Traweek 1988).

In the anthropological literature, encounters of this kind have received reflexive and critical attention. Although the meeting just described may seem neutral and innocuous, it has been argued that such meetings embed power relations and issues of epistemic authority. Thus, although the locals may be seen as the experts in their own cultural practices, the anthropologist is positioned as having authority to describe and analyze culture in a general form. In the words of Vincent Crapanzano (1992, 43), the ethnographer is, like the Greek messenger god Hermes, capable of first representing otherness in all its "opacity, foreignness, and meaninglessness" and then, through interpretation, penetrating the opaque, rendering the foreign familiar, and giving meaning to the meaningless: "He decodes the message. He interprets."

Yet in our case the distribution of authority was ambiguous, "lateralized" from the outset. For one thing, Brit was present at the performance as a partner with NatureAid Denmark. But NatureAid Denmark was there as

Figure 2.1
An ethnographer engaging in cultural performance.

a partner with NatureAid Vietnam, and both NatureAid Denmark and Brit were there as partners with the villagers who put on the "cultural performance."[5] Indeed, the establishment and strengthening of partnership was inscribed as one of the "immediate objectives" of the ecotourism project that was to be reviewed during the trip to the national park. As we will discuss in chapter 5, this review visit presented several "moments of partnership" in which all the issues we have pointed to—partnership as audit loops, as conflicting demands, as multiplied gatekeeping, and as divergent knowledge making—were made explicit.

Even in the photo, it is somewhat ambiguous who is in charge and precisely of what. The ethnographer is taken by the hand by a "cultural expert"—a young woman from the village—who clearly knows the choreography of the dance. But the privilege of representation resides with the NGO consultant, actually our "real" informant, who remains invisible behind the camera. Who represents and who is represented, and why and where, aren't certain at this moment.

One might think it pointless to put much emphasis on an evening of entertainment, insofar as the real task of the team we followed was to review and evaluate. Yet the Danish NGO members and their Vietnamese partners were constantly occupied with such "collateral" (Riles 2011) moments of partnership, and specifically with figuring out how such moments helped them to know the qualities of the project that was under review. Thus, dancing, trekking in the forest, and meetings were all "performance settings" (Heath 1998a, 71)—as much techniques for making knowledge as the formal indicators and statements that made up the final review report.

In these situations, then, our partners were our informants while they were themselves both informants and partners for each other. No wonder this posed "lateral" questions for all the actors involved in this development project—questions that included what counts as accountability and what as power; what counts as knowledge and information, indeed even what counts as infrastructure and partnership, when, and for whom.

Recursive Ethnography

In subsequent chapters we show how partnership and ideals of participation simultaneously provide context for and are embedded in a variety of existing and emergent organizational and technological practices occurring among NGOs, IT designers and programmers, and auditors. All these actors interrelate in different ways with existing and emergent aid information infrastructures. We show that imagining, designing, and working with aid infrastructures require simultaneous attention to, and handling of, meeting formats, indicators, databases, and information tools. As we will see, they also involve many new kinds of users and publics, some imagined and some rather more real. They also entail, we argue, the creation of new aid ontologies that are, to a significant extent, digitalized. Throughout the book, we trace the complex, recursive, layered relationships between the various kinds of actors that inhabit these infrastructures and practically shape these ontologies (Jensen 2010a).

We have suggested that partnership offers an interactive form whose strength and popularity are related to its capacity to encompass a very wide variety of settings. As the illustrations in previous sections indicate, partnerships cut across what usually are taken to be quite different kinds of sites (aid projects, technology development, academic conferences, performance audits) and quite different levels of analysis (global or local; micro, meso, or macro). This is why the recursivity of partnerships requires rethinking not only of the notion of analytical levels but also of the relation between the empirical and the analytical. If partnership takes place in situations in which empirical, theoretical, and analytical concerns interrelate, where does this leave us? If the empirical and the conceptual are recursively constituted in these settings, standard assumptions of who has political authority (donors over recipients, for example) and epistemic authority (project managers over villagers, or researchers over informants, for example) are challenged. This is why analytical approaches premised on structures of power and knowledge cannot be used to explain the effects of infrastructure and partnerships in development. Power and knowledge are themselves at stake in the recursive partnerships through which such infrastructures are made.

In view of the recursivity of partnership, of infrastructure, and of ethnography as a means of making knowledge, the analytical endeavor in which we are engaged in this book can therefore be characterized as partnership "seen twice" (Riles 2000, 23–70): as simultaneously practical accomplishment and as conceptual orientation. Partnership must be "seen twice" in the infrastructures we study because it encompasses our research while our research attempts to encompass it.

This situation has methodological implications of various sorts. It poses the question of how to delineate the "global" field of aid information infrastructure when each description is invariably both of a locality and created in negotiation with the partners inhabiting that locality. At the same time, however, it requires attentiveness to how certain infrastructural efforts create scale-changing effects and begin to become more global. Dealing with these two problems simultaneously is difficult precisely because while the infrastructures are emerging all the actors have different views of exactly how and why this is happening, at what level(s), and to what effects.

Commenting on the ethnographic challenge he faced in the Copper Belt of Zambia, James Ferguson asks:

> What happens to the anthropological understanding in a situation where "the natives" as well as the ethnographer lack a good understanding of what is going on around them? What if "the local people," like the anthropologist, feel out of place . . . ? The sharp line between the natives and the ethnographer, the locals and the foreigner, under such circumstances, becomes blurred. (1999, 19)

In spite of the empirical incommensurability between Ferguson's setting and ours, his characterization captures an important dimension of an ethnographic challenge we also faced. In our situation too, no one could be quite certain where the field began or ended, who was inside and outside, or what was important and what negligible. For all parties, the infrastructural space of monitoring movements remains emergent, fluid, uncertain, perhaps "spliced together" (Heath 1998b, 68), but certainly under construction.

Polymorphous Engagement

Multi-sited ethnography is George Marcus' (1995) well-known term for the methodological effort to handle increasingly mobile situations in which ethnographers are required by their informants not to stay in one place, because they are themselves constantly on the move. This certainly matches our experiences. After all, development practitioners are notorious travelers.

Moreover, many of the people and objects we wanted to follow were in places to which we could not gain access. (See Nader 1996.) Indeed, we were regularly reminded that "participant observation is a research technique that does not travel well up the social structure" (Gusterson 1997, 115–116). In the Danish Ministry of Foreign Affairs, for example, we were told that our methods would add nothing to the quantitative approaches already in use, and that consequently there was no reason to talk to us. Somewhat paradoxically, we struggled to be allowed into the very offices and organizations that envision or try to implement transparency and accountability measures.

Even so, problems with studying up are not the main reason why this book isn't based on multi-sited ethnography in Marcus' sense. We never had the ambition to follow particular people or objects around. "Without question," David Mosse writes (2005a, 18), "new challenges are presented to anthropologists of development by the 'real time contextual study of ideas and institutions' (Olivier de Sardan 2005) of elites and administrators, policy advisers, consultants or aid agency staff." We agree with this view. The dispersed settings in which infrastructure and partnerships co-evolve called for a somewhat different way of conceiving the field.

Faced with this challenge to our ethnographic imagination, we found Hugh Gusterson's methodological proposal for an ethnography centering on "polymorphous engagement" congenial. As Gusterson writes, polymorphous engagement involves "interacting with informants across a number of dispersed sites, not just in local communities, and sometimes in virtual form; and it means collecting data eclectically from a disparate array of sources in many different ways. Polymorphous engagement preserves the pragmatic amateurism that has characterized anthropological research, but displaces it away from a fetishistic obsession with participant observation." (1997, 116) In this vein, our methods were developed and fine-tuned in relation to specific sites. For some purposes, we have drawn on virtual ethnography (e.g., Beaulieu 2004; Hine 2005; Jensen 2010b). Thus, we view Web material, blogs, online newsletters, Skype meetings, "face-to-face" interaction, observations, and regular interviews as equally important ethnographical materials. In addition, we, too, left traces both in cyberspace (blog, list of conference participants) and in written project material (terms of reference, documents, copious email messages). Indeed, this was unavoidable, as we became participants in the field, and occasionally also partners.

We might point to a form of symmetry here. Just as our informants, through their multiple roles as partners, continually create their fields of interaction, we also had to invent ours in the very process of doing research (Jensen 2010b, 2012). Yet because our partnership relations were always under negotiation, we often had the sense that we were simultaneously "in the field," still "entering" it, and already in the process of "exiting."

One consequence of this polymorphous engagement is that our analysis doesn't rely on an unambiguous distinction that clearly separates the collection of data in the field from the subsequent theorization of ethnographic material at the desk. At no point did we encounter an obvious, stable, once-and-for-all distinction between field and desk. To claim that our analyses were based on a distinction between a field in which we were implicated and a desk at which we could "detach" would be to give our analyses an unwarranted epistemological advantage. Among other things, it would make invisible the consequential ways in which the field also localized and constructed our research orientations.

Starting from an exploratory field trip to Hanoi, the purpose of which was to negotiate access to a Danida project on Water and Rural Sanitation with the embassy, we were led to an odd assortment of other development actors, organizations, and institutions.[6] Those sites and actors were not directly empirically linked. Yet we now argue that they are in fact related, each a part of a global monitoring movement. They are related via their divergent engagements in varied activities through which partnership and infrastructures of accountability and transparency co-evolve.

Opportunities of Proximity

As we emphasized in the empirical vignettes at the beginning of this chapter, partnership is at once a practical, an empirical, and a conceptual issue. In her 2000 study of NGOs, Annelise Riles encountered people and organizations who used the same analytical vocabulary—that of the network—that she herself used. Riles wondered about the point of describing NGOs as a network if the actors themselves never ceased to conceptualize their activities using the same idiom.

Analogously, we studied partnership in contexts in which all the actors saw what they did as partnering, and therefore, as a matter of course, localized us as partners too. Riles suggests that in such cases analysis may seem stifled or even preempted, since everyone already knows what you want to be able to know. What stories are then left to tell? For what audiences? And what are the criteria of relevance?

In our case, central terms were shared with informants to a significant extent. They obsessed over partnerships, accountability, monitoring, and transparency; so did we. They hoped that our analyses would facilitate the improvement of what many considered suboptimal infrastructures and knowledge-sharing practices. Early on, we were surprised to see how easily many informants accepted our Latourian (1987) attempt to conceptualize their efforts as involving the construction of "chains of accountability." Later we realized that this metaphor (connoting an extended network, linked via a set of standards and other information intermediaries) was known and used in the field (although, not surprisingly, this usage was entirely shorn of actor-network analytical implications).[7]

At the level of terminology, at least, distance had literally collapsed. After initial frustrations, however, it became clear that the nominal identity of a shared vocabulary, as exemplified by "chain of accountability," covered over substantial differences in the scope, the meaning, and the implications of terms.[8] After this realization, this shared vocabulary became a generative source of friction that we could use to probe the distance and differences between our vocabularies and modes of knowledge making and those of different actors inhabiting the field.

Still, Riles' articulation of recursive relations between ethnographers and their informants is important in situations in which it *appears as if* conceptualization is shared. To us it has been useful in coming to terms with situations in which informants, having studied Pierre Bourdieu or Michael Power, felt quite comfortable with social scientific discourse, and were occasionally happy to correct not only our empirical mistakes but also what they saw as our analytical ones. (See chapters 5 and 6.)

In this context, Riles' discussion of "the Network" offered inspiration and helped us to specify practical ontology as a specific mode of engagement with the knowledge practices of development aid—and more generally. Focusing on "the Network" as both empirical and analytical in nature, Riles asked whether there was "indeed something new and disorienting about the forms of knowledge that people in the world of international bureaucratic practice worked and lived through? If so, what were its elements, how was it conceived, and what responses did it elicit? Who or what were the subjects and objects of such knowledge and what were its effects?" (2000, xiv) In the following chapters we are inspired by Riles' "solution" to the problem of recursivity: to treat the "the proximity of anthropological and technocratic knowledge practices" as "a kind of methodological opportunity" (2006a, 72). Notice the similarity between Riles' sense of "opportunity" and Deleuze's wasp and orchid, which share proximity only in order to go on diverging. Riles' and Deleuze's opportunities of proximity are both generated by a lateral move that encourages the anthropologist to take informants' knowledge practices as seriously as they take their actions. It means seeing such practices as offering conceptual challenge rather than merely empirical illustration (see also Jensen and Gad 2009; Gad and Jensen 2010), without, for all that, taking all of these views at face value. It also facilitates a rethinking of the transversal connections between the field of study and "epistemological and aesthetical commitments of one's own knowledge" (Riles 2006b, 16).

In this chapter we have sought to locate our own knowledge production amid the concerns of transparency and accountability held by the development actors whose practices we studied. In particular, our endeavor is to ontologize the work of informants and researchers *at once*. This has the peculiar implication that we consider empirical material and theoretical concepts, epistemological concerns, and observed practices as equally agential—that is, as components in building different kinds of worlds.

In the next chapter we begin to explore the inventive frontiers of aid infrastructure making. As always, we start *in media res*. Thus, the story is loosely structured as a travel narrative taking the reader from aid development practices in Vietnam to aid infrastructure development in the United Kingdom and the United States.

3 Inventive Frontiers: Aid Information Infrastructures and Their Users

Aid Information as a Complex Passage

Development aid is continuously changing its policies and concepts, yet many argue that it continues to fail to meet its goals—to build infrastructures, modernize, alleviate poverty, ensure gender inclusion and equity, and so forth. (See, e.g., Easterly 2006.) Why this is the case (and, indeed, whether this is the case) has been a subject of academic and policy discussions for several decades. At the same time, the perception that development aid fails to live up to its promises and expectations has become a driving force in the ongoing transformation of its practices.

For one thing, there is today a heightened (indeed a quite intense) focus on measuring the effectiveness of aid. If effectiveness can be measured, it will be possible to compare the successes and failures of aid across time, geography, and sectors, and to relate success or failure to administrative practices ("good governance") and political disposition ("democracy" or "dictatorship"). Yet, precisely because aid effectiveness is widely perceived to be elusive and hard to achieve, development specialists and politicians alike discuss what kind of information is needed to build a solid knowledge base on which future aid policy and practice can rely.

In this chapter we will explore several different (though partly overlapping) initiatives to make aid information publicly available. We begin with a story about how we first became interested in aid information during a visit to Hanoi. We then turn our attention to a conference on aid transparency and corruption, held in Oxford, that instantiated many of the concerns to which we were exposed in Vietnam.

Following the tradition in STS that has aimed to activate infrastructure by paying close attention to shifting configurations of technologies, people, organizations, and classifications (Beaulieu, de Rijcke, and van Heur 2012; Bowker and Star 1999; Suchman 2006), we pay particular attention to new forms of users that were articulated during the conference. Who, we ask, are the imagined users of aid information, and how do they come into being as relations are forged through aid information infrastructures?

Our starting premise is the non-neutrality of information. The sociologist Michael Halewood notes that "information is one element of the constitution of a subject." "However," he continues, "this is not limited to sense-perception (or reception); it is a complex passage of both qualitative and quantitative information that comprises the materiality of subjectivity." (2005, 82) The notion of information as a "complex passage" provides us with an initial guideline for interpreting the concerns expressed by the aid practitioners we met in Hanoi and Oxford. It inspires us to specify the emergence of new user or subject positions through infrastructure. And it obliges us to pay special attention to the complex passages and transformations between subjects and objects that shape emergent aid ontologies. Such ontologies are never fixed; their constituent entities are dynamic. They become what they are through processes of materialization and imagination, and through lots of work conducted by both humans and things. This complicates the widespread belief that achieving aid transparency is primarily a matter of designing the right technologies for the intended end users. Instead, it provides a way of coming to terms with the co-evolution of users and technologies.

Our story begins at a café in Hanoi, but the primary empirical material is drawn from a conference on Aid Transparency and Development held in 2010 in Oxford. In this chapter, the conference figures as something akin to what the anthropologist Sally Falk Moore has called a "diagnostic event" (1987, 730). The notion of a diagnostic event points to how a field site can function like a microcosm of larger issues, insofar as local affairs connect with social fields beyond the empirical setting. Moore expresses this as follows: "[L]ocal affairs cannot be addressed without serious attention to the larger processual implications of the local moment. Internal evidence of connections with social fields beyond the ethnographic site can lead to some of those historical significations." (1987, 735; see also Knorr Cetina and Bruegger 2002)

Though it is evocative, the notion of the diagnostic event doesn't fully capture what we are getting at. Amending Moore's use of the term, we suggest that "local" affairs might be addressed, not simply as examples of global processes, but as recursively and performatively implicated in such developments—as re-making the very scale of aid. When we elaborate on the Oxford conference, then, it is because it functioned as a conceptual and empirical "event" through which issues and problems relating to aid transparency were condensed and articulated as extendable forms.

To begin, however, we look back on our first visit to Hanoi, where the relation between aid effectiveness and aid information infrastructures was initially brought to our attention.

From a Development Assistance Database to Publish What You Fund, via Hanoi

We were introduced to some of the practical and political complexities of aid effectiveness during a conversation with an experienced Australian development aid consultant over a cup of coffee in Hanoi. In practice, we were told, mutual accountability—one of the five principles of effective aid specified in the Paris Declaration—"broke down to one meaningless indicator,"[1] because the notion of mutual accountability is "very political."[2] In spite of the discourse on mutuality, she explained, donors relate to their multiple stakeholders in quite asymmetrical ways. Since donors are, first and foremost, accountable to their taxpayers, demonstration of aid effectiveness at home is invariably most important. Mutual accountability might require new collaborative forms to have a realistic chance to improve aid effectiveness. Yet in most cases dramatic changes will be difficult to bring about because of the asymmetric accountabilities involved.

We had traveled to Hanoi to conduct interviews at the Danish embassy, and with the embassy's Vietnamese government partners, about how accountability relations are established and embedded in development infrastructures and organizational relations. Only days before, we had been told about an entity called a development assistance database (DAD)—a Web-based tool that the United Nations Development Programme had purchased. Developed by Synergy International Systems, this database was said to be running in multiple countries. It had been put into operation by the UN office in Hanoi. The DAD, we were told, enabled donors to report summaries of planned and current aid projects and to make available detailed information on the funding they provide for projects. The purpose was to enable recipient countries to use this information as the basis for budgeting and planning of activities. All of this was made possible by a website to which information could be uploaded in real time.

On the homepage of the OECD Development Aid Committee (www.oecd.org/dac/), the database is described as a flexible and easily configurable system that can be tailored to the needs of different countries. A similar point of view was taken by a young UNDP professional whom we interviewed in Hanoi. He characterized the DAD as consisting of a "core DAD," webpages "based on DAD technology," and webpages "going beyond the DAD technology." The purpose of the system, he said, was to facilitate the sharing of knowledge about program activities and budgets, enabling donors to plan ahead and to think strategically about how to create synergy among their initiatives while offering recipients instantaneous information about disbursement plans. However, that appeared to be rather far from current uses of the DAD. At the embassy, employees said that information continued to be missing or outdated, and was patchy at best. Even when information was available, its validity was questionable. We were told that data from certain donors systematically looked "too good," presenting a too precise—and therefore unrealistic—correspondence between original budgets and actual expenditures and showing too much progress toward programs' objectives. The embassy employees made it clear that they wouldn't trust the DAD as long as donors known to be inefficient and uncooperative were able to present information that made their operations look flawless. Using information technology to ensure mutual accountability seemed a challenging task.

Our conversation with the Australian consultant corroborated this view. Emphasizing the importance of aid effectiveness, she provided additional information about the sociopolitical basis of these difficulties. In addition, she made us aware that a development assistance database isn't simply *a* development assistance database. First, there isn't only one such technology; there are several. Some of them build on others; some compete. Some platforms are actual; some are imagined; several are under construction. Second, it is difficult to understand these different databases as merely technological. Rather, their existence or development is bound up with multiple, particular configurations of actors, organizations, and platforms, and with real and imagined uses, users, and publics. Thus, aid effectiveness was described to us as intrinsically linked with the emergence of a variably configured landscape of aid information infrastructures embodying different notions of accountability and knowledge. Similar to the characterization of infrastructure offered by Edwards, Bowker, Jackson, and Williams (2009), existing aid infrastructure seemed to be at once "always already there" and "an unfinished work in progress."

The consultant pointed us in the direction of an initiative called Publish What You Fund,[3] which she saw as a particularly promising way of moving the aid effectiveness agenda forward.

The Publish What You Fund Campaign and the International Aid Transparency Initiative

The London-based Publish What You Fund campaign was launched in 2008 at the Third High Level Forum on Aid Effectiveness, held in Accra. According to its homepage, it "represents the culmination of an initiative by two organisations—Tiri and Access Info—working in the integrity reform and access to information fields (respectively), who formed a coalition with a number of development-orientated CSOs (Action Aid, UKAN, and ONE) to establish the Publish What You Fund principles."[4] The founder of Publish What You Fund, and its director until 2012, was Judith Olson, an English development specialist with years of experience working with the United Nations High Commissioner in Croatia, the Rwandan ministry of finance and agriculture, and the English Overseas Development Institute. At the ODI she had worked on topics related to the behavior of donors and the effectiveness of aid. Immediately before starting Publish What You Fund, Olson had served as policy manager for the ONE campaign, a grassroots initiative against poverty and diseases. These varied experiences had gradually led her to the realization that any kind of practical development initiative would be inhibited by the general lack of transparent information on aid.

Whereas the DAD originated as an initiative of the UN Development Programme and was later distributed to recipient governments, Publish What You Fund defines itself in relation to NGOs and other organizations pursuing the goal of promoting transparency in development cooperation. In line with this aim, PWYF refers to itself as a campaign. Advocacy in policy circles is one of its main activities, the point being to change "the policy environment and policy debate." PWYF thus positions itself as a watchdog monitoring the operations of US and EU aid agencies.

One of Publish What You Fund's central activities is promoting and guiding the International Aid Transparency Initiative, which also was launched during the Accra meeting. PWYF aims to "ensure that the standards, formats and definitions developed by the International Aid Transparency Initiative fully reflect the Publish What You Fund principles." IATI works to define a standard (meaning a technical platform) that donors and aid recipients can use to display aid information. IATI doesn't work toward developing a comprehensive infrastructure for aid information. Rather, it aims to enable donors and recipients to report using one common standard. IATI employees refer to this work as developing a data registry rather than a database. IATI wants to change the reporting practices of donors more than it wants to change the databases. It hopes to do this by creating "a platform that enables a variety of intermediaries to use and present the information in different ways, including mixing it with other kinds of information, to develop new and existing information services (e.g. a partner country management system, a Uganda health portal, a budget tracking service, a website for parliamentarians that monitor new projects)."[5]

The IATI standard thus involves a "viewer" and a comprehensive system of coding and classification that breaks the codes used in the existing "gold standard" for aid data—the OECD-DAC's Creditor Reporting System—into smaller units. IATI presents itself as "a new multi-stakeholder initiative, which aims to increase the availability and accessibility of information about aid." By March 2010, 18 countries and aid organizations, including Denmark, Australia, and a number of private aid foundations, had signed the IATI principles. By August 2012, the principles had been endorsed by 22 countries, and 10 countries were represented on the IATI steering committee. IATI employees with whom we spoke expressed confidence that they would succeed in holding stakeholders to their promise of working toward more transparency by means of a pillory strategy—that is, by ranking stakeholders according to how much information they were making publicly available.

If the ultimate aim of IATI is to build a successful standard and ensure buy-in from the major donors, the ultimate aim of Publish What You Fund is to work itself out of existence by facilitating the mainstreaming of the IATI standard. This will have occurred when the campaign has reached its goal of having a robust set of transparency monitoring processes implemented in the practices of donors, recipients, and civil society.[6] Mainstreaming, again, involves changing the practices of data collection and reporting and making all information available for public scrutiny.

As these descriptions suggest, the target groups of the Publish What You Fund initiative are much more diffuse and expansive than those related to the development assistance database we encountered in Vietnam. As we have noted, PWYF aims to be recognized at the policy level and wants the United States and the European Union to sign its transparency principles. But the constituency is far broader. Under the headline "Who needs more and better aid transparency," PWYF's homepage outlined a situation in which citizens, civil society, and parliamentarians in both donor countries and recipient countries will benefit along with recipient governments (offices, auditing institutions, line ministries), private bodies, and NGOs.[7] On its website, the campaign thus suggested that implementing transparency measures would lead to a kind of win-win-win-win situation in which donors, recipients, civil society, and private businesses would all stand to gain simultaneously.

In the attempt to further PWYF's ambition, activities related to the making of evidence on aid effectiveness are folded into activities that seek to reach the public and create awareness of the need for transparency. As an organization that uses data to monitor how well policy is being implemented, PWYF aims to create a globally engaged public by connecting the interests of very diverse groups of people and organizations.

This comprehensive agenda explains why high-level policy advocacy is combined on PWYF's webpage with information activities that aim to reach out to a diffuse public of concerned citizens. The outreach aspiration is evident in an introductory video that explains the goals of the campaign in accessible language,[8] in the publication of a newsletter, and in invitations to contribute.

We continued to wonder about the specifics of those contributions. Exactly who would contribute what to whom, and why? Exactly what should it be possible to do by means of the data registry that IATI works toward developing and PWYF lobbies for countries and development organizations to sign?

The AidData Conference

Our questions continued to linger as we received a listserv notification about a conference on aid transparency and development finance. Since the notice specified academics and development practitioners as the intended audience, the event seemed relevant, even though at the time we didn't recognize the organizers, the AidData team. Hosted at University College, Oxford, the event was supported by the Global Economic Governance project of the college and the AidInfo Consortium. It would feature the launch of AidData, the "result of a merger between the PLAID (Project-Level Aid) database and the Aida (Accessible information on Development Activities) database of the Development Gateway." AidData, the conference invitation explained, "tracks more than 1 million projects spanning 81 donor agencies and totaling more than $4.1 trillion dollars over 40 years."[9]

As we were to find out, AidData is part of a network of organizations that seek to make aid information more transparent by raising awareness, through infrastructure design, or both. Based in the United States, it has the Bill & Melinda Gates Foundation and the William and Flora Hewlett Foundation as two of its sponsors. The conference of which we had been notified was funded by a wealthy American philanthropist who was keenly interested in fighting corruption.

For the conference, the AidData research team had invited actors interested in aid and transparency to give presentations on how to make aid more effective by enabling financial flows to be traced. The venue was the Examination Schools building at University College, a location that gave rise to comments from several speakers about stressful memories about exams taken in the same place. We mention this because it is indicative of the shared elite background of many participants now working in development organizations worldwide.

An introductory panel featured statements on the relationship between aid effectiveness, transparency, and corruption by researchers from the Global Governance Program, the Brookings Institution, Development Initiatives, the Center for Global Development, and a seemingly token speaker representing the Development Assistance Coordination Office of Sierra Leone. All identified aid effectiveness as crucial in combating corruption.

The early part of the conference was dedicated to exploring how AidData information could be used to change policy practices. Most of the speakers were from the AidData team, which consisted of researchers from Brigham Young University and the College of William and Mary, both in the United States. There were also representatives from the World Bank Development Gateway and from Development Initiatives.

One of AidData's principal investigators prefaced the public release of the database by introducing it as a public good. It was said to constitute a public good because it enabled any potentially interested person or organization to make free use of data. This is relevant for researchers, but also for makers of development policy and recipients of development aid; it enables them to compare and analyze aid flows from multiple sources, aggregating or disaggregating data using the more than 700 project codes in the database. This free public availability was among AidData's more distinctive features. Equally important, however, was the fact that the database featured the most detailed and comprehensive information on aid currently available, spanning the period from 1970 to 2010.

Using the previously mentioned "gold standard," the OECD-DAC Creditor Reporting System, AidData had modified sector codes to better disaggregate aid flows and to allow for the classification of projects with multiple purposes. In addition to all the information that could be dug up about "non-traditional donor" flows from China and other countries, AidData researchers had made available more detailed descriptions of aid projects wherever that was possible. In explaining why AidData should be considered a public good, the principal investigator emphasized that the data had been cleaned "in order to make it useful."

The second part of the conference was dedicated to exploring AidData. After a series of early presentations on the practical challenges of collecting, validating, and standardizing data, the organizers stated that the boring part of the conference had ended and the remaining time would be devoted to presentations based on actual use of AidData information. In these sessions, which had such titles as "Corruption, transparency and good governance," "Non-DAC donors—how much do we really know?" and "Donor patterns and aid quality," AidData's usefulness would be demonstrated. In presentations, economists and political scientists used AidData for various comparative purposes, visualizing aid flows using the available tools and offering copious statistical analyses. The analyses generated considerable excitement among many of the participants, especially among researchers interested in the quantitative analysis of financial flows and their relations to the effectiveness of aid.

Enacting the Ordinary Citizen-User

However, not all participants showed equally strong interest in AidData as a research application. One of the keynote speakers explained the importance of building "an evidence base" and using "technology to enable aid transparency." He repeatedly reminded the audience of the importance of figuring out what information would be most useful from a "southern user perspective." This speaker, an important figure in the transparency movement, was a master of telling stories that linked political visions for aid information and transparency with examples from the everyday lives of "normal citizens" in developing countries. He told the story of a village in Senegal where weather forecasts were read into a microphone to be broadcast through loudspeakers all the way to the coast, so that fishermen could use the forecasts to plan their activities in "real time." The point was that data as such is uninteresting. The question is, rather, what people are capable of doing with it. As the speaker explained, it is "the ability to mash up data that gives it strength."

The users invoked in this scenario were not economists or political scientists. Rather, they were recipients on the ground—citizens of developing countries who were supposed to benefit from aid. The main point of the presentation was that the key to success was having faith in the ability of users to do things with information, to feed back information, and to help improve the system. During a conversation outside the conference room, the speaker used an analogy to the United Kingdom's parliamentary expenses scandal when explaining his vision of how users with access to fine-grained information about development projects could become influential in ensuring the quality of the projects.

The parliamentary expenses scandal began to come to light when the *Daily Telegraph* made so-called called expenses files[10] available to its readers and arranged a competition to find errors in British politicians' self-reporting. As our conversation partner explained it, making fine-grained public information available to ordinary responsible citizens led to the uncovering of numerous irregularities in the use of public money by members of Parliament. The competition eventually led to criminal investigations that resulted in new regulation of politicians' allowances.

The parallel to the parliamentary expenses scandal conveys a clear statement of the kind of accountability relations that the transparency advocates who attended the Oxford conference would like to see embedded in aid information infrastructures. Accountability, in this perspective, relates to making detailed aid information public in the recipient countries, and also in donor countries, so that ordinary citizens can be properly informed. This virtual public consisting of citizen-users will thus be able to take on the role of responsible watchdogs capable of using "mashed-up data" to discipline governments.

Geocoding and the Wired-up User on the Ground

The organizers of the Oxford conference had, in fact, also given thought to non-research users. But, as they openly admitted, they found the question of who to involve through AidData, and exactly how, difficult to answer. This difficulty is probably what came across in the previously mentioned description of the newly launched database as a rather abstract, generic public good. Similar to the views expressed by the keynote speaker, AidData researchers also saw information made public as *in principle* always preferable to information controlled or hidden. Freed information makes it possible to establish feedback loops between a broad but unspecified set of aid providers and recipients. The precise feedback mechanisms, however, were not obvious to the researchers. Even so, the presentations made by AidData researchers invoked two additional user groups. We refer to these as, respectively, *wired-up users on the ground* and *not-yet-informatted*[11] *bureaucrats.*

To understand the imagined role of wired-up users on the ground, we must first consider some of the activities put in motion by the AidData team in order to engage this group. Among these activities, the team had organized "aid information challenge barcamps" in London and in Washington. These events invited the participation of development enthusiasts with a vivid technical imagination. The visions of these people (most of them young, well-educated Westerners) were tapped into at these intensive "barcamps," where ideas were generated and applications for mobile phones developed with which to make ingenious use of newly available aid data. In particular, inventive new ways of using geocoding were discussed. Geocoding involves mapping development projects in geographical space by means of geographical information systems (GIS). Rather than associating only a set of numbers (representing aid flow) with a set of codes (representing sector and types of aid activities) and a short description, the addition of geocodes to project information would allow for new ways of visualizing aid flows. In addition, it was proposed that geocoding could facilitate new ways of validating aid work in practice. This is where the wired-up users on the ground in developing countries enter the picture.

In a vivid example, one researcher explained the importance of geocoding for involving these potential users and thereby making them part of the AidData public. The scenario that was outlined involved an aid project in which a health facility was to be built in an African country. The AidData file would hold information that showed how much money was supposed to have been spent on this facility. It would also offer geocoding information on exactly where the building was supposed to be located. All this information could be easily downloaded to a mobile phone by means of an app developed at the barcamp. In the example, an African villager would download the app to his mobile phone and receive information about the local project. Then, as he bicycled past the location specified by the geocoding, he could take a photograph, which might show a well-functioning facility, an empty shed, or nothing at all. He could then upload this information to the file, empirically documenting the relative efficiency of the aid. Thus, local users would be able to force policy makers (also updated) to know that something had gone awry, or that a development success had been achieved. In this situation, the unresolved problem was how to align the new technological opportunities with the organizational, cultural, and economic opportunities to get recipients of aid interested in contributing to the AidData files. As we will see, it also had to do with creating aid professionals who would routinely check those files, then use information about missing facilities (uploaded by means of intelligent apps) to put pressure on policy makers.

Crossing the Street, Connecting Agendas: From AidData to IATI and Back Again

Members of the AidData team joked about the hard work of collecting information from a large variety of sources. The expression "brute force" was used repeatedly to capture the labor of "collecting data on the ground." With manifest excitement, one young researcher told us how he had traveled to different countries and explored their "legacy systems" to find ways of extracting data. In a memorable phrase, he recounted how he had "sucked data right out of the guts of the Czech Republic." Forty researchers and coders had busied themselves for approximately six years re-coding data from the OECD-DAC Creditor Reporting System and complementing it with data from numerous other sources, including donors' websites, evaluation reports, newspaper articles, and meetings with aid agencies. "Brute force," they explained, is "the best we currently have."

Yet while the members of the team clearly were proud of what they had managed to accomplish through "brute force," they were also aware that technologically their solution was far from ideal. In several panel presentations, the AidData database and the method of "brute force" were contrasted with the "elegant solution" toward which IATI was working by developing a "spine"—a "generic backbone of government activities at a global level" (Moon and Williamson 2010) for donors and recipients to use in coding and classifying information so as to make information on aid flows available to administrative bodies in recipient countries.

One striking difference between the current AidData database and the IATI standard is that the aim of the latter is to reduce the overall reporting time spent by donors by making reporting an integral part of regular administrative registration work instead of additional, or second-order, office work. In contrast, the AidData database builds on and supplements the Creditor Reporting System maintained by OECD-DAC. Donors in OECD countries are required to report to this system, but only every other year. The data is carefully validated and is acknowledged to be of high quality. Because it is retrospective, however, the data is of little use to ministries of budgeting and planning in developing countries that need real-time information on aid flows. Researchers, too, complained that the data in the OECD-DAC database was too limited and too difficult to manipulate, juxtapose, and compare. The elegant IATI solution would complement this system by constructing a far more flexible registry that would be useful to administrations worldwide.

While these discussions were taking place, we overheard conversations in which it was mentioned that several participants had business to do "across the street." What was so important on the other side of Oxford High Street? As we found out, the point of interest was an IATI-organized workshop being held elwewhere at the university. When we asked who the participants "across the street" were, we were told "The bureaucrats."

The venue of the IATI workshop was much more humble than the Examination Schools. IATI stakeholders congregated in small, dusty classrooms to discuss technical issues and issues relating to use of the IATI standard: Were user guides available? How could a link to the IATI standard be placed on the webpage of a ministry? In one classroom, "bureaucrats" discussed "user motivation." Their concern was how users might be motivated to do the extra work that would (initially) be required to retrieve data from IATI. What incentives were there to use the new system? How could IATI improve those incentives?

Whereas most of the participants in the AidData conference were from the United States or the United Kingdom, at the IATI workshop we found ourselves among participants from public administrations of African, Asian, and Latin American countries. End users from developing countries, generally absent from the AidData conference, were everywhere at the IATI workshop. This, however, didn't make it any easier for the aid transparency representatives to resolve the many issues that stood in the way of implementation of the IATI goals. Indeed, the presence of these users became something of a hindrance as they took turns asking perplexed questions about the relations between this technological platform and the practices and constraints of the everyday working routines of their home ministries.

The IATI workshop was over the next day. As the representatives of IATI and Publish What You Fund went back to the AidData conference, it appeared that most of the "bureaucrats" had left; at least their presence wasn't noticeable in the conference rooms. At the same time, the movements of aid transparency advocates from the conference to the workshop and back again suggest that, even though transparency infrastructure initiatives are separated by organizational affiliation and physical location (in this case Oxford High Street was the literal boundary), certain actors are involved in several partly competing and partly overlapping initiatives at once. Yet while the aid information infrastructures that these organizations work to realize and their imagined and actual users might be seen as evolving together, they converge only to a limited extent.

Inhabitants of the Inventive Frontier

One reason why the AidData conference can be viewed as a "diagnostic event" is that it simultaneously represented and performed the agenda of achieving aid effectiveness through data accessibility and transparency.

We have described a number of situations in which end users of new infrastructures of accountability were enacted. These users, and the publics they were seen to instantiate, emerged—discursively and, to an extent, practically—in the context of presenting new aid information infrastructures at

Oxford. In this section, we aim to further characterize these emergent entities. In doing so, we point to differences between users (real and imagined), and to the difficulty of developing information infrastructures that inscribe such a multiple (or multiplied) user group.

As we have suggested, one kind of implied user articulated at the conference can be characterized as the "ordinary citizen." (Cf. Callon, Lascoumes, and Barthe 2011, 122.) This typically Western user engages in creative interaction with aid information infrastructures provided by the organizations and initiatives we have described. The importance of these activities hinges on the expectation that new aid information infrastructures will provide information that will be relevant to a vastly diverse group of users. There is a shared assumption among the organizations we have described that more and different users are always better. When AidData is designated as a public good, and when the potential of making aid data available is elicited with reference to the parliamentary expenses scandal, the figure of an "ordinary citizen" willing and able to make innovative use of new Web-based data is evoked.

The various organizations that participate in the aid transparency movement expect that enrolling more users in aid transparency will add yet another win to the multiple-win situation that aid information infrastructures are hoped to create. This inclusive approach to the imagined and actual users is one of the components that turn the transparency agenda into a shared project despite the quite different aspirations, approaches, and technologies brought to the table by the different participants. The Publish What You Fund campaign is specifically articulated as a kind of informational consciousness raising in which all kinds of people and groups will be able to tell what is really going on with development funds. Yet whether the ordinary citizen is an entity that currently exists in the form required by aid information infrastructures is an open question. Indeed, the efforts to develop new information spaces seem, to a significant extent, to also be about creating new engaged publics that will be made up of citizens with an interest in development aid and a global mindset. Attempts to build these publics rely to a considerable degree on informational devices such as the AidData database.

We term a second type of aid information infrastructure user the *wired-up user on the ground*. Like the ordinary citizens, this group represents an emerging new public that engages in development policy by actively contributing to the transparency agenda. The user, dependent on geocoded project information, is exemplified by the bicycling African uploading aid information from his mobile phone. Obviously, the configuration of this user is based on certain unquestioned beliefs concerning the cultural practices of mobile phones, including individual ownership and exclusive use rights, which apply only to some parts of Europe and the United States. In addition, it seems that the wired-up user on the ground embodies an understanding of Web-2.0-ish technologies as the best and perhaps the only way to create aid transparency. Yet even if this user is admittedly the most uncertain and difficult to articulate for involved organizations, he or she is also—for the same reason—the best illustration of the inventive frontier of aid infrastructures.[12] We use the notion of an inventive frontier because the infrastructures are located at the very edge of existing aid ontologies, where not-yet-developed technologies meet not-yet-existent users. Uncertainty arises precisely because no one knows whether the passage between these technologies and users will be too complex to actually cross.

In stark contrast with these inhabitants of the frontiers of aid technology stands a third type of user, perceived in our context as living in infrastructural backwaters. We refer to these half-existing, half-envisioned users as *not-yet-informatted bureaucrats*. As we have noted, these users were represented at the IATI workshop that took place across the street from the AidData conference. There Latin American, African, and Asian government officials gathered to discuss the benefits of the new IATI "data spine" through which information about development aid activities could be reported and shared more efficiently. Although some of these not-yet-informatted bureaucrats also ventured across the street in the other direction, listening in on the AidData discussions, they were hardly present in the discussions there. However, they were very present in the classrooms where the IATI workshop was held. There IATI and PWFY representatives strove to explain the purpose of the reporting system they were developing and to ensure buy-in, interest, and sites for piloting their data registry. Meanwhile, not-quite informatted bureaucrats struggled to understand the technical details and concrete implications that the registry might have for their administrative work and for their relationships with donors.

An interesting paradox can be articulated through the contrast between the two sides of the street and the two corresponding types of users. AidData was described as a functional but inelegant technical solution, in contrast with the elegant but visionary IATI registry. The purely imagined users of AidData, however, were presented in a visionary mode, as offering elegant solutions to problems of aid transparency. Meanwhile, the users of the IATI registry were in a sense too real. Their very real bodily presence and the many questions they raised about how the registry would change their work and how it would help their administrations provided challenges to the designers of the elegant solution. The not-yet-informatted but actually present bureaucrat at the IATI workshop thus appeared as a kind of structural inversion of the already wired-up user on the ground, who was discursively present at but otherwise absent from the AidData conference.

We can amplify this contrast by noting that one likely reason why not-yet-informatted bureaucrats played only a marginal role in AidData visions was that *de facto* the bureaucracies of developing countries don't necessarily share the liberal vision of information as a public good that guided the infrastructural imagination of the project. To illustrate this point, let us return briefly to the development assistance database (DAD) that we encountered in Vietnam—the database that originally spurred our interest in aid information infrastructures. At first, it was acquired, sponsored, and hosted by the UN's office in Hanoi, which allowed free public access to the information it contained. Late in 2008, however, it was transferred to the Ministry of Planning and Finance as part of the ownership agenda. Since then the webpage has been transparent to the public only in a literal sense: one can view the content of the front page, as through a transparent film. One cannot, however, access any data or otherwise interact with it.[13] As this literally transparent source of information suggests, governments may insert their own opacities into the transparency agenda.

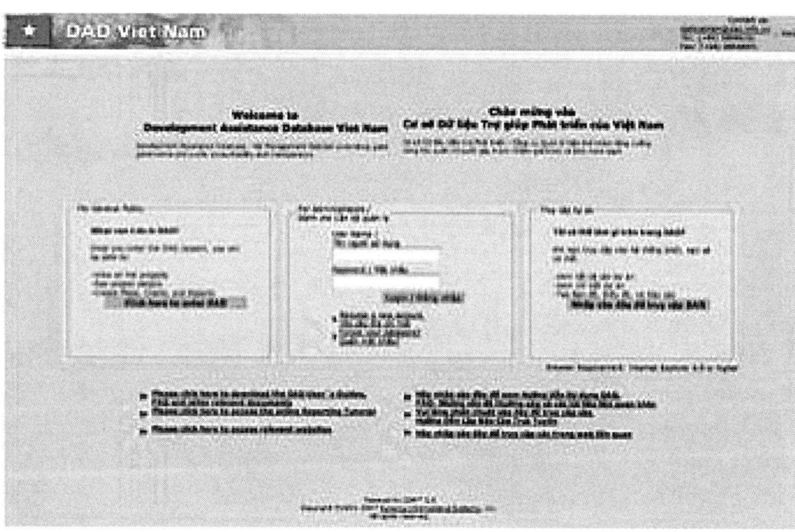

Figure 3.1
Literal transparency.

As a final user type, simultaneously central and tangential, we point to the *responsible researcher.* This user type is tangential because most of the development action to which AidData claims to be able to contribute has little to do with research. Yet responsible researchers are indispensable for the formation of this aid infrastructure. Indeed, without the researchers on the AidData team, it would not exist.

At the AidData conference, the predominance of academic presentations, analyses, and interests, in a setting that claimed to be about development practice, aid transparency, and corruption, was noteworthy. This is neither a critical nor an ironical observation. After all, the organizers, who had worked for years to create better sets of data on development aid, were political scientists. During their work they had gradually come into contact with the Development Gateway, the Gates Foundation, and other powerful and wealthy actors, including the philanthropist funding the conference. They seemed genuinely excited about the prospect that development practitioners and policy makers might use AidData for their own purposes. They hoped that the presence of representatives from governmental agencies and grassroots organizations would help bridge the gap between "research" and "practice." Indeed, some of the attendees were frustrated by the overwhelming academic focus of the conference. "I did not spend the last six years for this," one of the principal investigators whispered when yet another researcher embarked on an analysis of general trends in African aid expenditures. As responsible researchers, these people aimed to construct a quite specific outcome in which their research interests and the database these interests had led them to develop would tie in with broader development concerns. Even so, they had difficulty imagining precisely how this could occur.

Redistributing Aid Expertise

The relationship between aid information and its users allows us to pinpoint some crucial features of the ontologies of aid infrastructure development. The strongly felt need for new aid information infrastructure emerges in tandem with the proliferation of aid stakeholders; in turn, new users and publics come to be seen as important contributors and consumers of information. (See Winthereik, Johansen, and Strand 2008.) In addition, we have indicated how aid information infrastructures bear on questions about the relation between expertise and accountability. One reason why development aid practices must be made more transparent is to enable constituents to hold donors accountable to what they have promised. This is an effort to distribute information in order to enable a diverse set of users to gain knowledge about and expertise in matters related to aid. Yet precisely how this democratic redistribution of expertise may effect change through information remains uncertain. It depends on information categories and practices, and on the infrastructures that facilitate the collection and sharing of information. Thus, emerging aid information infrastructures raise questions about who gets configured as a producer of knowledge and who as a recipient, who can define the categories, who must live with them, and who is envisioned as a user but is rarely reached (Jensen 2010b). As we will continue to discuss in chapter 4, current efforts to improve aid transparency and accountability by means of information infrastructures have an important political dimension.

Co-Evolution, or Progression?

Our analysis has suggested that aid information infrastructures co-evolve with a varied, emerging set of users and publics. These publics are presumed to gain new interests and motivations, not least because of the new information made available to them. Thus, the making of aid information infrastructures entails imaginative efforts to invent new publics that will contribute to the creation of transparency and to new futures for development aid.

Of course this is not the language of IATI and AidData developers, who tend to talk of aid information infrastructures as moving up a hierarchy of technological sophistication and maturity. In this imaginary, some users are always more advanced than available technologies, and the most advanced technologies will thus be adopted by users eagerly waiting to be informatted or wired up. In this sense, a hierarchy of technologies corresponds with a set of progressively more sophisticated users.

Although everyone talked about the OECD-DAC Creditor Reporting System as the current "gold standard," it is also seen by advocates of aid transparency as located at the bottom of the ladder of information-sharing tools. It is in that humble position because, although it publishes high-quality data, the data is retrospective, static, and limited to formal aid documentation. In conversations with the participants in the Oxford conference, the OECD-DAC was discussed as a tool exclusively useful for the bureaucrat, whose expertise in handling information is limited to comparison of static datasets. DAD databases are seen to share the same deficiency, even though they encourage the sharing of real-time data among administrators at embassies, in ministries, and in donor organizations. The main reason for the perceived limitations of the DAD is that it is designed for governmental institutions and agencies and therefore doesn't radically "free" information by making it unconditionally available to the public and to users "on the ground" (as we saw exemplified by DAD-Vietnam's literally transparent interface).

A step up the hierarchy of transparency technologies, one finds the AidData database and the still-emerging IATI registry. From a technological point of view, there seemed to be unanimous agreement that IATI's elegant solution, which would merge routine registration, reporting, and publishing of data, represents an ideal for the future. Technologically that solution is at the peak of evolution. Yet it doesn't quite exist. Pragmatically, AidData's database, built on detective work to acquire information from "nontraditional donors," on the extraction of data from the guts of national legacy systems, and on the methodology of "brute force," is the current pinnacle of aid information infrastructures. Yet it struggles to find use for development administrative purposes, and it is largely unknown to its virtual addressees, the wired-up users in developing countries.

These differences also explain why AidData and IATI currently speak to radically different groups of users. AidData is built by and for responsible researchers who invoke ordinary citizens with a global mindset and wired-up users on the ground as potential new publics. Meanwhile, although Publish What You Fund makes much of the relevance of transparent aid data for global publics at large, the IATI registry is currently oriented toward bureaucrats, who have little immediate use for fancy apps and geocoding or for comparative research possibilities. While a researcher at the conference could say about AidData "the information is interesting, although I don't believe in it," the conference participant from Sierra Leone said that if data is not validated "we dare not use it for planning."

Finding information "interesting" although (or even because) it is "unbelievable" is an option neither for recipients of aid nor for aid bureaucrats, whose expertise depends on access to valid data. For the same reason, OECD-DAC representatives are skeptical of the idea that data collected from innumerable sources worldwide can form a useful knowledge platform. AidData and Publish What You Fund both argue the other way around, suggesting that, even though the initial data will be largely "shit," the transparency movement can really get going only by starting from the bottom of what we have called "the evolutionary hierarchy" and gradually working upward.

The differences in evaluating aid information infrastructures and in determining the relevant users indicate the limits of an evolutionary view of technology development. Using ethnography as a way of taking the efforts of Publish What You Fund and AidData seriously makes manifest that both the design and the use of aid information infrastructures are matters of bootstrapping to a significant extent. New technologies are built that embed user types said to exist in reality, while the building of these technologies also helps to conjure new publics that are going to have expectations about aid transparency. If AidData and IATI users are the imagined inhabitants of the inventive frontiers of aid information infrastructures, these inventions are never solely technological, but rather are technical, social, and political all at once. (See Suchman 2011.) If these initiatives are important for the agenda of aid transparency, it is not so much because of the solutions they currently offer as because their efforts turn information infrastructures into a self-propelling future-generating device (Jensen 2010a) for the practical remaking of existing aid ontologies. Thus, futures of aid infrastructure are continuously elicited through the evocation of half-real, half-imagined figures that are abstract enough for development agencies, development researchers, and non-governmental organizations to gather around without requiring them to commit to convergent plans for how to bring them into existence.

Thus, also, it is through "tangled networks of aestheticized information" (Orr 2000, 152) that AidData and IATI engage in creating new technologies and new users and in imagining new global publics. In the next chapter, we look more closely at some characteristics of the technologies that are meant to facilitate these transformations. We also examine some of the "goods" they promise to deliver. These technologies provide us with an entry point for exploring how aid information infrastructures rewire relations between technology and politics and for articulating some possible implications of these new designs.

4 Development Loop: Technological Politics for Transparency

Aid Information and "the Political"

In the preceding chapter we characterized the AidData conference in Oxford as a diagnostic event. It offered us an entry point from which to begin tracing the contours of the aid transparency movement, and it allowed us to characterize its emerging publics. Among the half-imagined, half-real users making up this constituency we identified several figures, among them the ordinary citizen and the wired-up user. We also pointed to the vividly felt presence of not-yet-informatted bureaucrats. Common among these types is a shared temporal orientation: all are users and are headed toward a future of increasingly transparent information. (See Jensen 2010a, 31–51.) Whereas the versions of the ordinary citizen and the wired-up user created by the technologists and the aid developers are in some sense quite abstract, universal (Oudshoorn et al. 2004), or potential figures, the informatted bureaucrat is rather more actual. People working in the central administrations of many developing countries are well aware of the demand for, and the potential of, making processes more transparent. However, bureaucrats are also tied up with everyday office work in ways that make their relation to the future of aid infrastructure ambiguous. As they work with existing systems and routines, their expectations for revolutionary technological changes is more circumspect than other visions we have discussed. A particular sense of guarded realism makes these bureaucrats embrace the promises of transparent information less fully. To come to terms with this discrepancy, this chapter explores some of the processes through which new aid infrastructure come to be imagined by its developers.

In physics, the term *potential energy* designates "the energy difference between the energy of an object in a given position and its energy at a reference position."[1] Wikipedia offers a simple illustration: "Think of a roller coaster. When the coaster climbs a hill it has potential energy. At the very top of the hill is its maximum potential energy. When the car speeds down the hill potential energy turns into kinetic. Kinetic energy is greatest at the bottom." Analogously, we argue that the energy invested in envisioned aid information users is maximized as potential in development contexts in which it has had little chance to be depotentialized as "kinetic energy." Such depotentialization occurs, we suggest, in the frictional movements whereby (as technology developers typically put it) aid infrastructures are "put into practice."

The limitation of this analogy is that the conversion of potential energy into movement sounds conspicuously like the "diffusion model" of innovation. (See, e.g., Rogers 2003. For a critique, see Latour 1987.) It encourages us to imagine new technologies diffusing into diverse practice owing to their own inherent momentum (their "potential energy"), unless hindered by specific "resistances" that generate friction. Indeed, this doesn't seem too far from how technology optimists view the matter. This picture is quickly complicated by the fact that, as in Latour's critique of diffusion, each movement of technology entails translation and transformation. When viewed in this way, friction isn't simply a matter of accounting for the energy of a system that dissipates through movement. Rather, friction is simultaneously what slows diffusion and what moves infrastructure in unforeseen directions.

Below, we further characterize some ways in which the development and the implementation of infrastructures of accountability are based on the imaginative, technological, organizational, and political skills of aid transparency architects and coders. In the previous chapter we paid attention to the configuration of particular kinds of users. In this chapter we will focus on how the potentials of new tools for transparency are presented online. We begin by describing the "demo" version of a user application for personal computers and mobile phones meant to facilitate transparency. We couple this discussion with an analysis of the political efficacy these technologies are imagined to have. The empirical base of this chapter includes both face-to-face interaction with people in the aid transparency movement and promotional materials found on YouTube that explain the design intentions and rationales of new apps for transparency.

Like other informational tools that aim to connect citizens and the state through Web interfaces (for example, health portals; see Adams, de Bont, and Berg 2006; Blobel et al. 2010), transparency apps do not make their public arrival fully formed. Of course these apps must be designed and programmed. But they also have to be enacted as technological tools that make it possible to address aid problems in new ways. This is why one important way of understanding these technologies is through designers' efforts to present and promote them. (See Winthereik, Johannsen, and Strand 2008.)

Promotional materials are particularly interesting because they allow us to study the "scripts" that are built into transparency technologies as they are made and as they are presented to the public.[2] At the same time, the political agenda of the aid transparency movement takes shape around what is practically possible in technological terms. Of course, designers are likely to imagine many things to be technically possible. Nevertheless, specific design efforts face constraints. New apps must work with existing databases and archives, which are not easy to change. We are thus witnessing a situation in which the design and development of aid information infrastructure and modes of political agency are co-produced (Jasanoff 2004). In particular, the case discussed below exemplifies how new technological forms can even induce transformations of what we think of as the political (Mouffe 2005, 8). It also allows us to characterize the digital ontologies of aid infrastructures.

Tools of Transparency

A news item in the March 2011 AidData newsletter makes it clear that one important characteristic of the aid transparency movement is the design of information tools:

> At their meeting in Paris on February 9, the IATI Steering Committee achieved a milestone in the aid transparency movement—finalization of a new global aid reporting standard. Jean-Louis Sarbib, CEO of Development Gateway, hailed IATI's achievement and noted that "this takes us a big step closer to getting complete information on aid activities to the communities where they are taking place, so that citizens are empowered to participate in their own development."

The newsletter announces that a milestone has been achieved with the launch of a new reporting standard. But the September 2011 issue of the same newsletter presents other kinds of tools:

> As the World Bank continues its push toward greater transparency, it has unveiled a number of improvements to data.worldbank.org. It's now easier to create charts, graphs, and maps using all of the World Development Indicators data. Techies will be enthused about the new embeddable widgets and API query builder.

Creating user applications that "chart, graph, and map" is one important way of approaching the transparency movement's goal of empowering citizens to "participate in their own development" through information. Apps, it would seem, embed an information politics.

In a 1998 book titled *Activists Beyond Borders: Advocacy Networks in International Politics*, the political scientists Margaret Keck and Kathryn Sikkink offer a number of observations that are relevant to this information politics. Advocacy networks can be likened to the aid transparency movement in that both work to promote global policy changes by means of globally dispersed information. (See Castells 1997; Chakravarty and Sarikakis 2006; Terranova 2004.)

Keck and Sikkink describe information politics as "the ability to quickly and credibly generate politically usable information and move it to where it will have the most impact" (1998, 16). They note that transnational advocacy networks tend to view such efforts as relatively inconsequential (since, as they say, "talk is cheap"). Nevertheless, they also point out that "once a government has publicly committed itself to a principle . . . networks can use . . . their command of information, to expose the distance between discourse and practice" (24).

Information politics is one of four advocacy tactics described by Keck and Sikkink. (The others are symbolic politics, leverage politics, and accountability politics.) However, while these other types may also be found at play in the transparency movement, there is no question that information is at the very forefront of its endeavors. For one thing, building an information infrastructure is the central aim of the aid transparency movement. As we have seen, as long as no transparent infrastructure exists, the achievement of accountability is assumed to be impossible.

In the settings we describe, therefore, the word *information* doesn't simply refer to one type of politics among others. Information is, rather, a prioritized category, seen to provide the basis for any kind of viable (that is, accountable and transparent) aid politics. Because information politics is embedded in aid infrastructures that are under construction, our focus is necessarily directed toward the making of these technologies and the technical and political imagination that guides them. This requires an inversion that foregrounds infrastructural form rather than political substance (Bowker 1994b; Jensen 2008).

We were first introduced to the Development Loop during the AidData conference, where it exposed us to the geocoding and mapping techniques that the AidData team saw as pivotal to the realization of aid transparency and accountability. In this context, the Development Loop was presented as a dream just about to come true. Intended to work on personal computers and mobile phones, the Development Loop exemplified how existing aid information databases such as the World Bank's data catalog (http://data.worldbank.org/data-catalog) might find practical use. Broadly characterized, the ambition is to create novel information loops between various nonhuman entities (such as the archives of the World Bank) and people (such as project planners, decision makers, and "aid beneficiaries.").

In a YouTube presentation of the Development Loop,[3] information and feedback figure centrally in enabling users (later characterized as project managers, planners, and citizens) to "begin to explore and ask better questions." Through information and feedback, users can begin to hold donors and governments accountable for their decisions and to evaluate development results, and "such questions should help donors allocate aid more effectively."

The particular information loop presented in the aforementioned YouTube presentation is brought into being as information about development projects travels from the World Bank archives to aid workers and recipients, moves into donor organizations, and eventually returns to development databases as new information. In the process, the promotional material argues, information will be exchanged, interpreted, and adjusted, and will eventually loop back into the archives in an updated and improved version. The archive will thus disseminate information to the users of the Development Loop. In turn, they will feed the archive through a process that the transparency movement hopes will enhance knowledge about aid practices and improve the foundation for making informed decisions about funding.

The YouTube demo and the website material make no reference to the institutional arrangements of which they are part. One might certainly question whether apps such as the Development Loop are able to work around the political inertia of powerful organizations such as the World Bank. (Cf. Suchman 1999.) But although this is an interesting topic, in the present discussion our focus is elsewhere. Specifically, we are interested in tracing some of the discourses and imaginative repertoires of people working to "activate" data archives, and, through them, aid recipients and development project workers.

In contrast to celebratory studies that view new infrastructures as having the potential to radically reconfigure current forms and values of knowledge production, Beaulieu, de Rijcke, and van Heur (2012) emphasize the importance of attending to the constitutive tension between socio-technical reproduction and innovation. Similarly, in this chapter we will explore some new knowledge configurations that are emerging at the interface of development practice and research, at locations where infrastructure is in the making. Adding to the analysis by Beaulieu et al., we argue that such knowledge configurations and their politics are outcomes of ontological transformations of the aid landscape itself.

Composing the Development Loop

According to promotional materials, the strength of the development loop is that "maps and layers" make it possible for users to visualize development aid activities at the project level. Users can compare information and can comment on what they learn as they do so.

Let us therefore proceed to look more closely at the demo of the Development Loop.[4] It shows a zoomable map with project dots similar to location markers in Google Maps. For example, one can search for project activities in a specific region and view all projects by the World Bank on water and sanitation in a particular region of Kenya. By adding more layers to the map, the user can compare the number of World Bank projects with the number of projects by other donors. Or one might compare donor presence to socioeconomic factors in the region by adding "a poverty map." This possibility is explicated in the demo with the example of yet another layer: a "heat map" that visualizes the intensity of investments and individual projects in a given area using a graded color scheme. The overarching purpose, again, is to allow easy (because visualized) comparison of donors' performances.

One of the Development Loop's particularly interesting features is crowdsourcing. By accessing information provided by aid workers and "beneficiaries," project holders (for example, donors and government agencies) can get a more nuanced picture of the progress of projects. The demo argues that this feature makes the Development Loop particularly useful for citizen empowerment. In a voice-over, this is explained as follows:

> Development loop is a Web and mobile application that tells the real story of foreign assistance and allows users to map and manage development data from the World Bank and African Development Bank.[5]

The promotional material promises that wired-up users, "even in low or no band-width situations," are enabled to influence far-away donors and governments by linking personal stories and experiences to a project on the map, or even by contributing to the accuracy of aid information by re-coding project information.[6] Using the Development Loop, the otherwise disempowered "southern user" can make "local" information seen "from the ground" visible in powerful decision-making forums. Thus, the demo video suggests that the Development Loop enables users on the ground to literally speak truth to power.

The YouTube demo video talks as if the primary users are people involved in project planning and decision making in the field and development professionals based in organizational headquarters. Nevertheless, aid

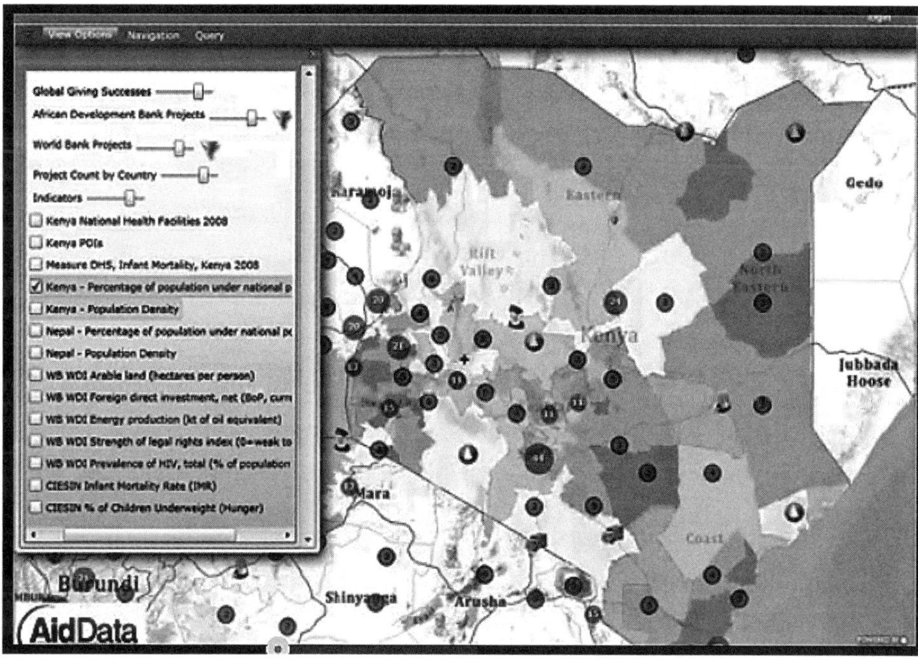

Figure 4.1
A screen dump from the AidData Mapping Prototype depicting a geographical area with multiple different projects, layered with markers of socioeconomic status.

beneficiaries are also invoked, though in more abstract form. And, indeed, the democratic potential seen to inhere in the Development Loop is manifest not least in the way that anyone, regardless of physical location or job title, can, in principle, contribute information (comments, pictures, numbers, stories) on individual projects. One way the progress of such projects can be viewed is by comparing project information with indicators based on the UN's Millennium Development Goals.

As an example, the YouTube video demonstrates that a map of all active World Bank project sites layered with "a poverty map" shows that the active projects are not located in the poorest areas: "This piece of information enables users to begin asking much better questions, such as whether aid is being allocated to those who may need it most."[7] By combining information from the World Bank with demographic information, the Development Loop thus promises to activate information held by the World Bank. In order to address the issue of aid effectiveness (that is, the suspicion that aid stakeholders don't know which aid projects and practices work and why), the Development Loop promises to combine information sources and to make them dynamic, thereby creating a better foundation of aid information and, ultimately, practical aid knowledge. These efforts exemplify and respond to what Geoffrey Bowker has called "the drive to database" (2005, 109).

Activating the Archive

The specific techno-political imagination on which the Development Loop draws is part of broader transparency tendencies. In 2010 the World Bank launched a competition called Apps for Development. Although the Development Loop was "not eligible for awards due to its maker being a World Bank vendor,"[8] it was listed as the 104th of 107 competing apps.

The announcement of the three winners of the Apps for Development competition emphasized that the point of the competition was to challenge the public to "create innovative software applications that move us a step closer toward solving some of the world's most pressing problems." A jury of tech gurus, including "Kannan Pashupathy of Google, Ory Okolloh, co-founder of Ushahidi, and Craig Newmark of Craigslist,"[9] chose the winning apps. According to the success criteria of the competition, the applications with a chance of winning would move us closer to solving aid development problems through information. These applications would do so, in particular, by making use of one or more of the datasets listed in the World Bank Data Catalog. In the same process, however, the competition also provided a way of raising awareness about this catalog, seen to be chock-full of underutilized information and only recently made available to the public. On the webpage, the World Bank explains the rationale behind opening up this dataset:

> The World Bank's Open Data initiative is intended to provide all users with access to World Bank data. The data catalog is a listing of available World Bank datasets, include databases, pre-formatted tables and reports. Each of the listings includes a description of the data set and a direct link to that set. Where possible, the databases are linked directly to a selection screen to allow users to select the countries, indicators, and years they would like to search. Those search results can be exported in different formats. Users can also choose to download the entire database directly from the catalog.[10]

When the catalog went online, in the spring of 2010, the accompanying press statement, signed by the World Bank's president, Bob Zoellick, emphasized the importance of free information. In making the point, it referred to the importance of accessible aid data for rebuilding devastated

Haiti. It also made a more general assertion: "Our experiences show that what gets measured can get changed."[11] In a commentary quoted on the blog and "personal democracy forum" TechPresident, the Swedish public health expert and "data visionary"[12] Hans Rosling concurred with that estimation, asserting that freeing the enormous World Bank Data Catalog would "foster innovation."[13] It was precisely to tap into this innovative potential that the opening up of the catalog was followed by a call for the submission of creative apps.

The World Bank Data Catalog comprises an array of data collections, from World Bank Finances and World Development Indicators to Migration and Remittances Factbooks, Country Profiles, and Populations Rankings. The standard search function categorizes data collections as "global," "micro data," "regional," "research," "specialized," "tables," "topical," "World Bank operational," "finances," or "commodities." "At a glance tables" covering all countries "allow for quick overview over key indicators tracing the trends in social and economic indicators from the last three decades."

This is all to the good, but a problem (one we have encountered before) crops up. The problem, as seen by advocates of aid transparency infrastructure, is that, although the catalog contains vast amounts of data, it is static. The World Bank has an enormous amount of information about aid projects in its hands, but so far it has had few ways of activating this information as "living" knowledge. How will such activation take place, and for whom will the knowledge be "living"? The apps that were submitted to the Apps for Development contest provided concrete answers to the first of these questions—answers that also required implicit or explicit consideration of the second question.

As a passive receptacle for information, the World Bank's data catalog provides no tools for using its data for dynamic and comparative purposes. However, apps such as StatPlanet World Bank (the winner of the competition) and the Development Loop promise to do so. They would make it possible to gain "feedback from the ground level" and combine it with information from the archive. Yet linking up the World Bank's data catalog with new apps takes work. In particular, project information has to be geocoded in order for catalog data to be accessible in the Development Loop.

The project that geocoded the information for the Development Loop app was called Mapping for Results (probably a play on the Paris Declaration's "Managing for Development Results"). That project, too, was conducted as a partnership—or, indeed, as a "development loop"—of Aid Data university researchers, the Development Gateway, and the World Bank.[14]

In an entertaining YouTube video that presents the ambition behind the coding efforts, some young programmers doing the manual geocoding are interviewed about their efforts.[15] Accompanied by Chopin's *Heroic Polonaise*, they introduce the practice and the ideology of geocoding by explaining that it is about rendering "aid flow geography" visible by "talking with graphics" in such a way that the locations of aid projects are made transparent to the user. Such information, we are told, isn't easily accessible through existing search engines and data collections. Even though one is potentially drowning in *general* information about aid, *specific, location-based* data is very hard to come by.

As we argued in chapter 3, information inevitably entails complex passages that reshape organizations and users. To make more detailed aid information publicly available is, thus, not a neutral activity. These efforts can be viewed, for example, in the context of sociological and historical studies, which have documented how modern state making is intimately connected with developing infrastructures that allow states to classify and know their citizens (Curtis 2002; Desrosiéres 1998; Scott 1998).

Such knowledge facilitates new possibilities for managing and intervening in personal and public affairs. The availability of new information can effect scale change and ontological transformation: what used to be a local issue or an individual problem becomes connectible with broader social and political issues, or changes such issues altogether. This is precisely how Helen Verran describes the capacities of numbers: "As routines of gestures and utterances, ritualized repetitions, numbers negotiate a scale shift, from the minutiae of family life to the macrosocial category of British Empire." (2001, 100) Considering the complex passage whereby location-based information promises to create relations between a bicycling African aid beneficiary and a World Bank archive exemplifies such scale-changing capability *in potentia*. We further view it as illustrative of politics ontologically reconfigured through aid information.

Still, even though our interlocutors were implicitly engaged in "politics," they saw questions of classification and information granularity as primarily practical. Transparency advocates with whom we spoke had no political program other than the "meta-program" of freeing information. Thus, they simply designed solutions that would offer aid organizations and grassroots activists a better knowledge base for their development aims. Accordingly, we never heard mention of the idea that one might design infrastructure for the empowerment of particular groups or subject positions. Nor were these design efforts accompanied by any very specific considerations of who would come to inhabit the positions made available by apps such as the Development Loop. But, of course, configurations of users occur regardless of reflexive consideration or deliberate intention.

The way the programmers tell it, we are witness to a classical story of knowledge asymmetries. (See Jensen 2010b.) Donors supposedly have location-based information, but so far they haven't made it available. No one really knows where aid activities are happening. "We" (the programmers speaking in the video) "don't know," and "they" (aid recipients, planners, NGOs) also "don't know" how projects are really doing. Thus, planners, implementers, and citizens are powerless as a result of their lack of knowledge. However, the coders explain, their geocoding efforts contribute to changing this state of ignorance.

In the YouTube video, the careful re-coding of the World Bank's data catalog and of other collected aid information takes place in a large open office. This "bunker location" is identified as 1889 F Street, the home address of the Development Gateway in Washington. Because Mapping for Results is a partnership between researchers and development practitioners, programmers as well as World Bank professionals give their perspectives on the benefits of geocoded information. In the video, a strong sense of solidarity and teamwork in face of the huge task of making the static archive active and useful in new contexts comes through vividly.

At one striking moment, a coder expresses her hope that in the future there will be "some sort of system" in which geocoded data will be collected routinely by donors and kept updated. Her statement indicates that maintaining an updated and thus "healthy" geocoded database is by no means a simple task, both because information is scattered across multiple sources and because aid projects are always changing. Also, it suggests that manually geocoding the World Bank's data catalog is a major task—in all likelihood a one-time event. Even though, according to the video, Mapping for Results managed to geocode 12,000 projects in seven weeks, there are hundreds of thousands of projects in existence. It will be a daunting task to code all the files, not to mention keeping them up to date. And it will be very difficult as long as the effort has to be carried out manually. It was probably not at random that the *Heroic Polonaise* was chosen as background music.

The problem articulated by the enthusiastic coder mentioned in the preceding paragraph is akin to an issue of maintenance that was highlighted by AidData's principal investigators at the Oxford conference. Attempting to convey the laboriousness of the "brute-force methodology" that the AidData team had used to "hand-code 115,000 OECD projects," and their manual efforts to "suck data out of the guts" of multiple incongruent national legacy systems, all with the purpose of integrating them in the

AidData database, one presentation likened this work to the heroic but ultimately futile struggle of the steel driver John Henry, who famously defeated a steam engine but died from the effort.[16] In view of the unsustainability of continued hand coding of projects, this conference presentation grappled with the question of whether "machine coding" might provide a viable solution. However, as the authors recognized, the trustworthiness and precision of their Bayesian classification algorithm was uncertain at best. They found no easy way out of their John Henrian predicament. It may be a long time before the Mapping for Result programmer's imagined "sort of system" will be able to replace manual geocoding and the mundane task of updating files.

Ontologies for Transparency

In previous sections, we described initiatives to turn cataloged aid information into dynamic knowledge of aid and of particular donors' effectiveness. In the eyes of our informants, these initiatives are currently seen to hold a lot of potential energy; a lot of actual energy is also currently spent to bring their promise to fruition. But how might this potential, its process of actualization, and its political implications be characterized? The situation is complicated by the fact that these development efforts draw simultaneously on the conceptual "energy repositories" of the notions of *information technology* and *transparency and accountability*.

Information technology is a term that, in our time, can usually be trusted to invoke connotations of futurity and progress (Robins and Webster 1999), as indeed it does among advocates of aid transparency. As regards aid accountability, it is certainly tempting to look for technologically mediated answers to aid problems in a context in which aid investment is broadly seen as having been inefficient for a long while. Apps like the Development Loop and projects like Mapping for Results tap into and draw on both registers of promise: political efficacy and fairness strengthened through technology made democratic and ubiquitous. If the imagination guiding these initiatives is oriented toward the future positive, how might social and cultural researchers engage seriously with their aspirations without swallowing them wholesale? In particular, how can they do so without using the critical tools of sociology and cultural theory to turn the people involved in these activities into dopes (Garfinkel 1967, 68–75) who don't know what they are really doing, since more sinister agencies (neoliberalism, capitalism) are working through them? Before pinpointing how practical ontology enables us to address these questions, we will briefly sketch two approaches that we do not find satisfying.

The first approach is oriented toward technological feasibility. For example, technology-oriented studies may critically examine the pros and cons of current efforts to create transparent aid infrastructures and evaluate their likelihood of success. Yet the broader social analytical potential of such studies is necessarily limited, since the most critical questions that can be raised from within this horizon of interpretation concern the technological or economical viability of these systems.

However, insofar as we are equipped with critical cultural and social theory, another sort of approach is even more tempting. We are thinking of the kind of study that would locate aid information infrastructures in a broader political landscape. With this ambition, one might ask whether these technologies are likely to have any political effects at all, and, if they do, whether such effects are likely to be positive, and for whom. It is plausible to argue, for example, that recent developments in aid infrastructure are not likely to change the dominant political ideology of aid or the general institutional structures of aid development (Bissio 2001; Ya'U 2005). Consequently, technologies such as the Development Loop are likely to be seen, ultimately, as offering old wines in new bottles. (Cf. Wong 2003.) Even if aid professionals get access to heat maps that indicate the ineffectiveness of current modes of operation, for example, they may be both unwilling and unable to change current practices dramatically.

In spite of the critical "good sense" of each of the two approaches we have sketched here, we nevertheless proceed in a different direction. We are not interested in judging the feasibility of these specific technologies, but we are equally unwilling to evaluate their potential on the basis of assumptions about what in aid's current political ideologies and structures can and cannot be changed. Instead, our vantage point is that the political implications and characteristics of aid infrastructure are only gradually elicited (Star and Ruhleder 1996, 32). This process generates the "slight surprise of action" (Latour 1999, 266–93) through which different versions of technology emerge in tandem with different versions of the political.

The political theorist Chantal Mouffe argues for the importance of distinguishing "politics" ("the manifold practices of conventional politics") from "the political" ("the way in which society is constituted") (2005, 8; cf. Barry 2001). Whereas Mouffe is primarily interested in intervening at the level of "politics," we are interested in using empirical material to trace emergent relations between infrastructures for transparency and "the political." Specifically, attending to the ontological work done by transparency infrastructures, we argue that they also contribute to giving form to new versions of the political—versions that are hard to grasp with conventional understandings of politics.

Politics and the Internet

To see how this might work, we turn to some theoretical arguments of current social studies of the Internet. Engaging these arguments is important for two interrelated reasons. On the one hand, it enables us to qualify some of the ways in which critical social theory has addressed the political implications of information infrastructure. We do so with reference to the aid transparency movement and the information tools we have introduced. In reverse, scrutinizing these initiatives in light of critical analyses of the relations between the Internet and politics allows us to sharpen our interpretation of the political implications that follow from changing development ontologies by means of tools for transparency. Specifically, we employ the notion of digital ontology to characterize aid infrastructure development as a distinct political space in what grand theorists have characterized as "the age of the Internet."

Up to this point we have stayed rather closely to an examination of how technological development and imagination are enfolded with projects of political mobilization in the specific context of making apps for transparency. In a more general sense, however, this case replicates discussions that are as old as the Internet itself. What we have in mind is the question of the Web as a site for political action.

In a 2007 book titled *The Internet Imaginaire,* the media historian Patrice Flichy offers a history of the Internet with a particular emphasis on the social imagination that undergirds notions of technological innovation. He describes how, in the mid 1990s, Vice-President Al Gore's expression "Information Highway" became the guiding metaphor for telecommunications liberalization, and how it later became affiliated with the promotion of democracy, development, and international peace at large. Flichy refers to this project as a utopia and a slogan in search of a mission:

> In a speech to the International Telecommunications Union (ITU) in Buenos Aires, [Gore] emphatically defined the role of information highways: "In a sense, the [Global information infrastructure] will be a metaphor for democracy itself. . . . The GII will not only be a metaphor for a functioning democracy, it will in fact promote the functioning of democracy by greatly enhancing the participation of citizens in decision-making." (2007, 31)

Flichy uses the trope of the information highway and its implied democratic potential to consider whether community practices facilitate the invigoration of organized political action on the Internet or whether, in contrast, it marks the end of politics. More specifically, the issue is whether new information technologies should be seen as harbingers of democratic progress (though they will have to be pragmatically adjusted to real situations) or whether they must be seen more critically as offering new tools with which to retain or strengthen old political and institutional agendas. The suggestion that the Internet era marks "the end of politics" indexes a strong version of the latter argument. It suggests that the simultaneously anonymous and individualized form of interaction that the Internet makes possible erodes the public sphere that is presupposed by "the political."

The generic argument for political reinvigoration is that the Internet provides an avenue for grassroots democracy, and that anonymity ensures that everyone online has an equal voice. Online, it is said, race, gender, and money count for nothing, because no one knows who you are. Yet, as the novelist Jonathan Franzen has remarked about "the digital revolution," "if both Newt Gingrich and Timothy Leary are crazy about it" then "somewhere, something isn't adding up" (2002, 170). Like the win-win-win-win situation outlined by Publish What You Fund, the view seems too happy to be true. It seems to "cosmetically" cover over too many political, technological, and practical differences to hold under closer scrutiny. Consequently, the counterargument comes as little surprise: "It is a mistake to think that the Internet is an inherently democratic institution or that it will necessarily lead to increased personal freedom and increased understanding between people." (Flichy 2007, 157) Flichy goes on to argue that the Internet cannot be viewed as a neutral space detached from other aspects of social life. His analysis pays attention to the fact that the Internet is both a product and a shaper of cultural, political, and economical practices. (See Abbate 1999; Hine 2000.) In a first estimation, therefore, his point is agnostic: the Internet, *as such*, neither hinders nor improves democracy. Although Flichy's language is different, this point can be readily translated into the terminology of science and technology studies: when considering the political implications of the Internet, we cannot accept *a priori* arguments either for or against its democratic potential, since whatever happens will be an effect of translations between multiple social and technical actors and practices. Indeed, the very suggestion that the democratic potential of the Internet can be evaluated in terms of pros and cons is problematic, since technologizing politics is likely to produce many different kinds of costs and benefits for many different kinds of actors. Rather than viewing the Internet as a homogeneous sphere with unequivocal political effects, that is, it is necessary to consider how it inscribes particular geographies of responsibility—in the plural (Akrich 1992, 207).

Some critical theorists, however, are not averse to making more sweeping diagnoses. The postmodern Marxist Arthur Kroker has pinpointed one consequence of the Internet as the establishment of a "virtual class." "The virtual class," Kroker argues (as quoted in Flichy 2007, 165), "is the class that comes to power on the back of cyberspace or the Internet." Kroker suggests that the virtual class operates on the basis of an ideology of facilitation. Franzen succinctly elucidates how this ideology operates: "*Wired* attempts to celebrate the in-ness of the in crowd while leaving the door open for newcomers, and it manages the trick by selling both vision and inside dope." (2002, 167)

According to these rather dire interpretations, the virtual class gains power by inviting non-users onto the Internet, invoking a liberal rhetoric of promise. Once online, however, these users are abandoned to discover for themselves that the Internet doesn't lead to any radical change in living conditions but is, rather, a continuation of the fierce economic struggle that they already know only too well from offline settings.

In broad agreement with Kroker's critique, Flichy argues that the Internet environment isn't conducive to the formation of a democratic utopia, since it is organized according to neoliberal capitalist principles. Internet users are interpellated (Althusser and Balibar 1970) not as citizens or political subjects but only in the circumscribed role of consumers, enabled merely to choose between products. Online, interaction is entirely optional, since "at any point in time citizen-consumers vote with their feet or, more exactly, their mouse" (Flichy 2007, 161). Since one can always choose to exit any interaction, the Internet is basically a space of non-commitment. Thus the Internet marks the end of politics.

Transposed to our analyses of the architects and coders of the aid transparency movement, this diagnosis is astoundingly recognizable. Even so, however, it is unable to account for several distinct features of this case. Flichy (2007, 161) quotes the law professor and legal activist Lawrence Lessig to the effect that the Internet "can be characterized by its 'architecture of liberty.'" If that is the case, one way of addressing the pertinence of the critical diagnosis is to consider how well it allows us to account for the activities of aid transparency programmers, who literally aim to redesign this architecture of liberty. Surely their efforts are infused with rhetoric of promise and aspiration. Significantly, however, "selling both vision and inside dope" is no straightforward task.

A Virtual Class?

Now, the resonance of Flichy's analysis with our case should be readily apparent. It isn't difficult to identify our programmers and development professionals as exemplars of a virtual class with explicit recourse to an ideology of facilitation. They attempt to mobilize aid recipients and project workers on the ground in order to challenge the existing power structures of development. And they might readily concur with the ambition expressed in the claim, made by John Perry Barlow, a co-founder of the Electronic Frontier Foundation, that "we are creating a world that all may enter without privilege or prejudice, accorded by race, economic power, military force, or station of birth" (quoted on p. 159 of Flichy 2007). The power of transparent information is indeed meant to induce such a situation. Facilitating comparisons between donor performances, for example, is seen as a first step toward a situation in which donors can be neutrally judged through the information they provide, and can be forced to change their practices by the risk of public shaming.

The resonance of this argument can be further explored by taking into account the suggestion that the Internet marks the end of politics by disconnecting users from the public sphere. It isn't far-fetched to see these programmers as exemplary of what Jon Katz calls a new elite inhabiting a "post-political" sphere. According to Katz, members of this elite share certain common values: "They tend to be libertarian, materialistic, tolerant, rational, technologically adept, disconnected from conventional political organization. . . . They are not politically correct, rejecting dogma in favor of sorting through issues individually, preferring discussions to platforms. . . . The digital young are revolutionaries. Unlike the clucking boomers, they are not talking revolution; they're making one." (quoted on p. 164 of Flichy 2007) According to Flichy, these are people with a "new view of politics" and the will to implement it: "[They] were not interested in swearing allegiance to a party that would more or less force its program onto them. On the contrary, they wanted to deal with politics rationally, like one sorts out a technical problem." (165)

Yet in spite of the obvious similarities between notions of a "post-political sphere" characterized by an "ideology of facilitation" and discourses of transparency through design, there are also reasons why our app builders can't be so neatly categorized as inhabitants of a virtual class. In conceptual terms, the overarching problem is that the critical interpretation offered by Flichy functions by cordoning off the domain of the technical from the political. However, this immediately leads to questions of empirical interpretation. Not least, it is far from easy to make sense of the suggestion that the Internet invariably undermines the possibilities for political action in a context in which technologies are built for explicitly transformational, though undefined, purposes. Indeed, accepting that view would require us to discount as irrelevant all the practices and sentiments we have come across so far.

What can be asked, of course, is how and why the sphere of political action is imagined and created as it is. Flichy's suggestion that politics is dealt with "rationally, like one sorts out a technical problem" is obviously pertinent in this context, in which information-technology architects deal with transparency and presumed deficits in democracy as informational issues to be solved through geocoding. But again we need to pause. What does Flichy assume about the technological when he worries that the political might be treated as a technical problem? Centrally, this involves the presupposition that technical problems are "rational" in the first place. That view, in turn, is based on the idea that technical problems are free of social and political context. Our case testifies to the more general STS insight that these domains co-evolve; that they mutually shape each other in unforeseeable ways.

Thus, the present case allows us to deploy the insight that technical problems, even if approached "rationally," invariably entail social relationships and processes. Programmers busy with manual geocoding of the World Bank's data catalog yearn for the day when "some kind of system" will allow for full automation, while aware of the unlikelihood of that scenario and deferring it to an indefinite future. Meanwhile, even as AidData research assistants recount their John Henrian efforts to hand-code OECD project files, they are also enabled by participating in this project to "pursue research interests of their own," to travel to conferences, and to "speak with corporate executives, diplomats, and area experts" (AidData monthly newsletter, September 2011). While "the political" in aid transparency initiatives is undeniably infected with a technical imagination, then, technical work comprises multiple collateral tasks (Riles 2011) and poses varied challenges: it requires artful problem solving and handling organizational relations, and it engages programmers in hard work that is professionally and personally gratifying.

Procedures of technical problem solving invariably also involve social interpretation. Located in Development Gateway offices, programmers never cease asking themselves whom they are addressing and how. Who might be the users that will pick up on these technologies? What kind of infrastructure will be needed in order for new kinds of publics to have a chance to emerge? These social questions are inseparable from the technical activities that engage aid transparency programmers.

In light of the situation just described, we can point to some specific flaws in Flichy's suggestion that the political is undermined by the technological. The argument presupposes the virtual class as a general and homogeneous category and is enabled to do so because of the broad characterization of this class as wholly individualized and technically oriented. Yet our case illustrates a situation in which individualized action is not opposed to building platforms (both literally and figuratively) because a technical orientation *entails* social embeddedness—and vice versa. Rather than people gradually developing "new views" on the use of technology, the co-evolution and mutual transformation of infrastructure, its designers, users, publics, and politics imply that we are in the realm of ontological transformation.

This leads to a second flaw in Flichy's diagnosis. By assuming a static notion of the political—a public sphere defined in opposition to an individual realm—critical studies such as Flichy's actively disallow us from taking seriously the activities of the aid transparency movement, which busies itself with redefining the political through information infrastructure. Given the limitations in existing social theoretical vocabularies, we have to develop analytical modes that allow us to grapple with potential transformations of the political by the technological without prejudging whether such change is good or bad.

Coded Presence

Our analysis suggests, then, that rumors about the end of politics in the age of the Internet are greatly exaggerated. Yet this doesn't mean that all is business as usual. Indeed, the notion of the political may be undergoing mutations due to changes in the infrastructure of aid. To the extent that existing democratic forms (for example, representative or deliberative) are seen as the only viable versions of the political, obviously it will seem as if "the political," as such, is threatened by new technological developments. There is, however, no reason to assume that. More interesting to explore is how the political co-evolves and transmutes in tandem with the design of infrastructures of transparency. Indeed, initiatives such as the Apps for Development contest and the activation of the World Bank's data catalog are illustrative of certain interesting vectors of political transformation.

In a 2006 book titled *Virtual Migration*, the anthropologist A. Aneesh discusses the changing conditions of the transnational labor market in light of the information revolution. Aneesh suggests that capitalism itself becomes refigured in a global context in which the rule of code comes to organize business practice. Aneesh describes collaborative, distributed work practices in which, through programming, "persons have been so fully integrated into the work process that it is hard to separate them from the process itself." In this sense, collaborative teams become literally part of "a *machine*." Aneesh emphasizes, however, that this is a novel development, and that it doesn't simply replicate or intensify the mechanistic horrors "satirized in the Charlie Chaplin movie *Modern Times*." The reason, similarly to how we have described the work done by programmers, is that in order for "the machine" to work efficiently, the rule of code becomes infiltrated with, and depends upon, such "nonrobotic elements" of interaction as "motivation, competition, and co-operation."

Still, if there is one strikingly common aspect to the activities we have described, it is precisely the way information—embedded in new technological platforms—is seen as a driver of political change. Aneesh's analysis alerts us to the fact that the social and political imaginations that guide the creation of tools such as the Development Loop depend on coded and distributed presence. Coded and distributed presence indexes a future situation in which policy makers and aid workers, via technology, can access real-time aid data from projects all over the world. Meanwhile, distributed aid recipients are offered "coded presence" in development agencies' headquarters through the information they upload. In this way, the rule of code provides a crucial translation mechanism whereby dead information about aid can be turned into lively and dynamic knowledge. Efficacious political action consequently requires active engagement with evolving aid information infrastructures. But since such engagement requires knowledge of and interest in the opportunities engendered by these infrastructures, as well as the capacity to engage with them, it is clear that this emergent "architecture of liberty" (Flichy 2007, 161) is radically different from the ideals of both representative and deliberate democracy.

Digital Ontologies and Political Action

In this chapter we have focused on some current activities in the aid transparency movement, a particular kind of social movement with a distinct techno-emancipatory project. In particular, we have attended to the Development Loop and a series of activities around it, the Apps for Development contest, the "opening" of the World Bank's data catalog, and the specific geocoding efforts of the Mapping for Results project. We found a community of programmers busily turning information from the World Bank archives into a dynamic dataset suitable for integration with the infrastructure they are busily designing and promoting. We also heard their reflections on the purposes and challenges of this work.

As has been noted, this project aims to empower globally distributed and wildly heterogeneous users by giving them access to aid information. The curiosity is that its aim is to enable political action through infrastructural development, irrespective of what the content of specific programs or campaigns might be. Were this infrastructure to gather momentum, information politics would indeed turn out to be politically important not just for advocates of aid transparency but for any transnational advocacy network. Whatever their specific topic, the political impact of these networks would depend on putting information collated from aid information infrastructures to use. In this case, as McLuhan (1997) would have it, the techno-political medium, in the shape of aid infrastructures, would become the primary message.

Of course, it remains to be seen what political effects such aid infrastructures will have. However, one plausible interpretation is that, just as non-governmental organizations are wary of information politics because "talk is cheap," coded information will become cheap in a situation of universally accessible aid data. Contrary to words, which are cheap because they don't necessarily lead to action, radically freed information may well cheapen because it delegates the problems of selection, combination, and interpretation, which have traditionally been handled within the context of particular organizations or projects, to any actor whatsoever who is interested in the information.

Rather than a shared foundation of knowledge, we surmise, this situation is likely to lead to a multiplication of disagreements, conflicting evaluations, and contested proposals. Rather than creating more homogeneous knowledge about aid that can be used to create more effective interventions, it is likely to render knowledge heterogeneous in new ways. (See Winthereik 2003.) As we will elaborate in chapter 7, infrastructural ontologies of transparency—and emerging aid information infrastructures in general—locate us within a monitoring movement in which individualized actors such as organizations and people become increasingly distributed, re-combinable, and partial—that is, become *dividuals* (Strathern 1988; Deleuze 1995).

The activities we have described are obviously illustrative neither of a full-fledged transparent aid information infrastructure nor of a radical ontological transformation. Aid information infrastructures do not yet have sufficient depth or width to work as they are intended to work, and no one would claim otherwise—least of all the advocates of transparency. Yet these are infrastructures that are trying hard to gain depth, reality, and consequentiality. (See Jensen 2010a, 19–31.) Gradually, through much work, simultaneously social and technical, they are transforming from pure potentiality to moving technologies in friction to changing aid realities. Viewed as ongoing processes of practical ontology, aid infrastructures continuously spin off utopian and dystopian visions for the future.

Yet while aid transparency advocates and technology builders are busily engaged in concretizing aid infrastructure, such infrastructures are already in existence elsewhere, though in quite different forms. For example, quite different versions of aid infrastructure, also meant to ensure particular kinds of accountability and transparency, are embedded in NGOs' multiple forms of partnerships. The next chapter provides a contrast to the digital ontologies of transparency described in this chapter as it takes us into the everyday, but no less complicated, infrastructures of accountability that employees of the NGO NatureAid already work with and against. These infrastructures, however, are also on the move. For that reason, they are also, inevitably, in friction.

5 Weedy Infrastructure: Monitoring Environmental Partnerships

Monitoring in Partnerships

In previous chapters we described some inventive initiatives that aim to create better platforms for knowledge about aid activities and projects—knowledge that is meant to be beneficial not only to aid organizations but also to policy makers and to various publics. A study of these endeavors, however, didn't get us in touch with any actual aid development projects. So far, we have heard little about how such projects operate, or about how accountability and transparency matter to the people who instigate them. And we have heard next to nothing about the infrastructures for accountability, transparency, and knowledge making that are already in place—the infrastructures that aid organizations currently rely on in their daily operations. Yet such infrastructures do exist, and, as we shall see, they too are inventive in their own ways. These ways, however, have little to do with cutting-edge technological innovation or macro-level knowledge sharing. Rather, such aid information infrastructures result in the invention of organizational and collaborative practices. As Frederick Cooper and Randall Packard say, "much development knowledge is down to earth" (1997a, 18).

Nevertheless, these kinds of practical inventions are also hugely important. They concern always urgent questions about how development consultants can gain knowledge about aid projects taking place all over the world and, centrally, how such knowledge can be transported back and forth between aid headquarters, donors' offices, and non-governmental organizations' project sites. In this context, infrastructure making and the knowledge practices that undergird it are themselves political. Indeed, infrastructure making is part of an ontological transformation that reshapes what the politics of development aid is and what it can be.

This chapter focuses on "monitoring and evaluation" ("M&E") in the Conservation Department of the Danish office of NatureAid. NatureAid Denmark has an executive board and a secretary general. It is divided into three departments: Conservation, Fundraising, and Communication. Located on the renovated third floor of an office building in a run-down part of central Copenhagen, the three departments share a large, semi-open office space. Our fieldwork centered on monitoring and evaluation activities in the self-described "Third World Team" of the Conservation Department.[1]

In recent years, NatureAid has been short on funds. In this situation, the Copenhagen office has worked to produce new forms of evidence of the effectiveness of its projects. Occasionally, such evidence can be brought to the table during negotiations with donors or other stakeholders. More often, it is simply part of everyday routines. Formal systems of representation are deeply embedded in this infrastructure, and standardized documentation and reporting formats are omnipresent in the office's work.

We entered NatureAid's office by signing a standardized form, called the "terms of reference," that had been drafted and revised through negotiations with the head of the Conservation Department. The terms defined our partnership with NatureAid and stipulated that we would be allowed to study its accountability practices for a year. In return, we would organize two workshops during which staff members would be invited to reflect on existing modes of "doing accountability" in the organization.

Anthropologists have argued for the importance of documentation practices in organizational life (Riles 2006c; Schwartzmann 1989). In our case, too, formal documentation was an unavoidable technology for knowing the realities of projects. For team members, documents provided important resources for coordinating, planning, and legitimizing project activities. In some ways, they even *produced* such realities. To illustrate how documents participate in the making of aid infrastructures, our analysis focuses on the making of a document: a midterm report written as a review of an ecotourism project in South Vietnam.

In spite of their importance, however, formal reports are only one element in the infrastructure of accountability that ties NatureAid Denmark and its many partners together. Studying the midterm review allows us to examine the complex processes whereby a fragile infrastructure is established between monitoring and evaluation in the formal worlds of aid bureaucracy and "in the wild" (Hutchins 1995) of aid projects.

Exploring the efforts to prepare and conduct the midterm review, we show that accountable aid infrastructure is a provisional outcome of collaborative processes, and that it is an element in the multifarious, ambivalent relations that make up partnership among Danish team members, Vietnamese NGO partners, local villagers, and others. Successful knowledge making is conceptualized by the Danish review team as a matter of successful enactment of participation and partnership. However, the Vietnamese partners have different ideas about what constitutes partnership and project success and about how formal and informal monitoring and evaluation activities might support either. In this context, partnership is simultaneously an important *precondition* of the review and an *outcome* that must be achieved if the review is to work at all—that is, both a means and an end.

In what follows, we show how partnership is enacted in the midterm review. This happens through a series of negotiations in both "formal" and "informal" settings. These various settings provide the techniques through which project knowledge is created and made visible and by means of which partnership emerges. We also attend to "moments of partnership," a term we use to refer to temporally, spatially, and relationally delimited situations in which issues of particular empirical and analytical interest emerge (Rod 2010; see also Corsin-Jiménez 2007 and Strathern 1999).

Similar to diagnostic events (see chapter 3), moments of partnership highlight the emergence of a defining aspect of a particular practice. Yet moments of partnership cannot be taken as condensed forms of more distributed activities. Instead, they remain inherently attached to the specific set of local interactions in which they emerge. In our case, the particular form of partnership emerging at these moments has implications for the role that formal project documentation comes to play in this fragile infrastructure of accountability.

Weeds and Cultivation

Researchers in both information systems (Hanseth n.d.) and the anthropology of development (Mosse 2005a) have proposed the notion of *cultivation* as a tool with which to analyze the complex relations between formal requirements and informal practices. But the notion of a cultivated alignment between formal and informal practices fails to do justice to the intricacies of infrastructural accountability. Rather than cultivation, we focus on the "weediness" of practices of partnership and infrastructure making. The sociologist Zygmunt Bauman uses weeds as an entry point for discussing social order:

> What makes some plants into 'weeds,' which we mercilessly poison and uproot, is their horrifying tendency to obliterate the boundary between our garden and wilderness. The 'weeds' are often quite nice looking, fragrant and pleasing; we would certainly admire them as adorable specimens of wildlife if we found them while walking through woods or a meadow. Their 'fault' is that they have come, uninvited, to a place which ought to be neatly cut into lawns, rose garden, vegetable plot and flower borders. They spoil the harmony we envisaged, they play havoc with our design. (Bauman 1990, 57)

In the case at hand, what gets to count as weedy, when, and for whom depend on complex positions within the aid project. This suggests that Bauman's diagnosis of the problems of weeds also has limits. Indeed, his characterization of weeds aligns with the metaphor of cultivation, since it assumes that for social order to obtain weeds must be removed.

Below, we explore several instances in which Danish ideas about how monitoring and knowledge making in partnership produce weeds for Vietnamese partners and vice versa. In those instances, however, weeds do not pose a problem that can be solved through cultivation. For although weeds are troublesome in the context of project accountabilities, they are also unavoidable and generative for knowledge making in the partnership. Indeed, they are regenerative of partnership itself.

The Organizational Environment

In some respects NatureAid is an unusual NGO. Neither founded by nor run by nature-loving political activists (Tsing 2005, 121–155), it is not known for participating in activism involving public protests. (See Keck and Sikkink 1998.) Indeed, in stark contrast with grassroots activism, it was established by a member of a European royal family and subsequently "exported" to other countries through that family's elite network. An individual or a company may become a member through donation. NatureAid increasingly relies on these sources of funding. Donating is a member's only activity; unlike many NGOs, NatureAid doesn't use members as voluntary staffers.

NatureAid Denmark is part of an international network. Employees of the Danish office are in relatively frequent contact with employees of NatureAid offices in other countries. Country offices pay a fee to the international headquarters in Geneva; in return, they are allowed to use NatureAid's highly branded logo. The international network circulates presentations, best practices, consultancy reports, annual status reports, and relevant research findings. Headquarters also obliges the country offices to work with certain international standards. Among other things, headquarters define and distribute "Key Performance Indicators" and guidelines for planning, monitoring, and evaluating projects. Yet in spite of these standardization efforts, daily work is conducted with considerable independence.

For NatureAid's Danish employees, the international network (often referred to simply as "the Network") is an important though not constantly felt presence. The Network consists of more or less distant colleagues with whom the Danish team shares certain professional interests relating to nature protection and the environment. Synergy arises in the Network from the fact that each country defines its own focus areas and has its own expertise. The Danish branch, for example, specializes in timber, palm oil, fisheries, and climate change; other branches have different areas of specialization. The Network is able to function as an organizational environment for knowledge sharing and professional exchanges because of these differences. According to the employees, the Network also brings value to daily work and solidifies NatureAid's reputation as a "knowledge-based organization."

In contrast to the Network, the international headquarters was perceived as rather removed from the action. In the eyes of staffers, mechanisms for monitoring and evaluation promoted by headquarters often have little practical use. Recently, for example, a special monitoring and evaluation unit was set up by headquarters with the purpose of standardizing reporting procedures and tracking the progress and the effectiveness of the various projects run by its large regional programs. According to staffers in the Danish office, however, this unit has a rather "desk-like approach." It knows very little idea about what makes M&E information useful on the ground (for example, in relation to the "adaptive project management" strategy advocated by Third World Team).[2] And indeed the Third World Team is distinctly *not* in favor of analysis based on desk research. On the contrary, its members view regular supervision visits as necessary supplements to the obligatory bi-annual technical progress reports. Such visits are indispensable both for generating knowledge about aid effectiveness and for actualizing partnerships.

Knowing Projects

Technical progress reports (TPRs) are one of many kinds of documentation practices that make up the system of representation through which NatureAid and its donors know projects. Other formats include bi-annual monitoring and evaluation reports, final reports, and midterm reviews. TPRs require project partners in developing countries to reflect on the progress of projects according to criteria specified in funding applications. They are updated in a so-called logical framework, which is referred to in all subsequent reporting on a project's progress.[3]

Although frequent self-reporting is seen by headquarters as crucial to the stimulation of "on-site reflection on how well a project is doing," these reports are seen by team members to be quite insufficient means of gathering knowledge about the realities of a project. According to the Third World Team, supervision visits help create more substantial knowledge. Further, this knowledge is not seen as a mere addition to what is formally required; it is, rather, viewed as an important precondition for creating the knowledge that funders demand. Even so, the knowledge gained from supervision visits must still be expressed in a language and a style suitable to the formal infrastructure of accountability through which NatureAid is connected with donors. This poses a challenge to NatureAid's monitoring and evaluation practices. Although in principle it is possible to cut and paste formulations from previous technical progress reports, it takes a lot of work to translate specific details and observations made during supervision visits into these forms. At the same time, it is vital important for team members that such translations are made. Only by making the realities of projects formally visible through documentation can they get approval to make adjustments to their projects. Since experience shows that such adjustments are almost always necessary, team members have to find ways to formally represent knowledge gained during field trips to funders and partners. Team members ask how spreadsheets that purport to contain information about all of NatureAid's projects in an entire region are supposed to satisfy their need for concrete information about individual projects.

Shifting Agendas and Modalities

It is increasingly important for NatureAid to work with governments and corporations. It does so in order to promote its two main goals: protecting threatened species and promoting sustainable production. For several years, the conservationist agenda has been channeled through a discourse of biodiversity. Hence, the central interest has been defined as the protection of threatened species, particularly "key species."[4] In recent years, however, this focus has been supplemented with a focus on how sustainable production and development "for people" might go hand in hand with biodiversity concerns. This shift in the Network's priorities has not occurred without contestation. However, the Danish office is firmly on the side of an expanded agenda that sees biodiversity and social equality as tightly entwined. According to Magnus, a staff member who has a degree in anthropology and who now is working as a regional manager, the approach to nature protection taken in the Danish office has changed: "When I was hired, I replaced a biologist. Hiring an anthropologist indicated a paradigm shift." Another staff member, Hanne, expresses the new orientation this way:

> The traditional way of working in the Network centers on animals. We make an elephant corridor and, if there are people around, we say that for us to help the elephant we'll need to help the people, too. Sustainable production and development has not been the core of our interest, but it is coming. Not least, because management wants it.

To these employees, working with policy makers on sustainable production is a much more effective way of protecting nature than conserving the habitats of particular species. As Hanne says, "encroachers will find a way into the protected areas anyway."

Aside from this recent change in conservation philosophy, staff members in the Danish office have also experienced changes in the way large donors view NGOs. (See Lewis and Kanji 2009.) Before 2008, NatureAid's activities were funded and run as individual projects. However, development agencies have increasingly sought to replace this way of giving aid with a new "modality" centered on funding larger programs. New aid modalities are closely tied to increasing investments in transparency infrastructures. As development funds are transferred to development partners in "fragile" bureaucracies, the need for procedures that ensure accountability and transparency has intensified.

Currently, Danida funds a substantial number of the projects run by the Conservation Department. This happens through a large program scheme focusing on sustainable production and socio-economic development in three regions. Since NatureAid Denmark has been successful in getting funding for a full program with activities in these regions, the "income" of the Conservation Department is ensured over a longer time period. Time is also freed up because the organization is, at least in principle, less busy applying for funds for smaller projects.

One implication of program funding is that the Conservation Department must work more closely with partners in the Network in order to ensure their ongoing commitment and their willingness to take ownership of program activities. As staff members describe this changing role, it is no longer a matter of managing single projects; now it is a matter of making sure that change occurs across a whole national sector. On various occasions, however, they also emphasize that this is a very difficult thing to do. Because of this difficulty, they must constantly temper their inclination to "micro-manage" the details of particular programs. This can be a strong temptation, however, since, according to members of the Danish team, they usually have much more project-management experience than their partners. Indeed, the requirements of partnership and ownership sometimes work against effective solutions, since they keep team members from offering the kinds of advice they could have offered as "experts."

The temptation to micro-manage is heightened because the Danish team alone, rather than the partners together, is ultimately accountable to the donor. Nevertheless, staff members emphasize that NatureAid Denmark is no longer the "implementing partner," and that they are there mainly to help partners "build capacity" so that *they* will be capable of implementing a project's objectives.

The partnership philosophy embedded in recent approaches to aid, such as support through program level funding or "untied" aid given directly to a country's treasury, go hand in hand with initiatives that are meant to help NatureAid enhance its role as a strategic partner. Before 2008, project managers were assigned to projects that were basically managed from Denmark. However, since the philosophy of the program approach is that project management should be delegated to on-site actors, staffers in the environmental NGO have increasingly acquired a double role as facilitators and controllers, with the occasional addition of providing "technical assistance."[5]

Within the NGO, one of the visible outcomes of the shift from project funding to program funding is the engagement by one staff member in a new organizational unit called the Network Initiative. Network Initiatives are meeting places for stakeholders in the various regions. They can be seen as symptomatic of the organizational need for a "layer" between "First World" and "Third World" offices. This need has emerged in the context of shifting aid modalities toward the program level, a re-orientation that requires constant negotiation and alignment of strategic interest among multiple partners.

A more local consequence of these changes is that staffers have made attempts to revitalize an internal M&E system.[6] The system was developed a few years ago as a tool for collective budgeting, planning, monitoring, and following up on project activities. The purpose was to supplement the official formats provided by the donors and by NatureAid's headquarters with a set of more open-ended questions relating to the annual project cycle. However, the system fell into oblivion during a recent period of dramatic budget cuts and firing of staff members.

The members of the Third World Team regret this situation, because they believe M&E ought to provide opportunities for internal communication, knowledge sharing, and learning. As they see it, management is often focused on "big wins," such as getting projects with commercial partners up and running. The downside is that insufficient attention is paid to the details that make projects successful. In this context, practices of monitoring and evaluation became themes of explicit reflection.

What Is Monitoring and Evaluation—and Where Is It?

As we mentioned above, our terms of reference specified that we would organize two workshops with staffers in the Conservation Department. Some months into our fieldwork, the head of the department requested our input for a large application for funding that was under preparation. As is usually the case, the application was to contain a section on M&E implementation. The team was uncertain about how to articulate its approach. Thus, the workshop was to provide a space in which the challenges of working within the existing M&E infrastructure could be explicitly discussed.

Framing this issue relevantly was difficult for us. Our observations in the office had largely corroborated findings by Paul Edwards (2003), Geoffrey Bowker and Susan Leigh Star (1999), and other scholars of infrastructure. These studies have emphasized both the ubiquity and the invisibility of modern infrastructures. Infrastructures tend to fade from view very easily, and they may be very difficult to get into view in the first place. This corresponded only too well to our experience. Indeed, we felt that our grasp of the accountability infrastructures was rather too tenuous for us to offer the firm advice that the head of the Conservation Department seemed to want.

On the one hand, we had no trouble recognizing that some kind of infrastructure that supported formal documentation practices was in place. On the other hand, we had significant trouble getting that infrastructure into clear perspective. It wasn't that we were entirely clueless. Rather, we felt as if we were glimpsing the infrastructure only intermittently. We knew, for example, that a shared database containing progress information of ongoing projects existed. At the time of our fieldwork, though, that database was used primarily to get contact information about people in the Network. We also knew that, in practice, staffers juggled different kinds of information through the flow of their work. Some of that information came from partners' self-reporting, some from Skype meetings and email conversations with partners, and some from supervision visits. These types of information drew on and were fed into a variety of formats and templates that were used for different monitoring and reporting purposes. Finally, we knew that, because no common system enabled staffers to get an overview over projects, translating between documents and formats in order to put together a coherent picture was a troublesome though crucial task.

Thus, monitoring and evaluation seemed to be happening constantly, but not at any particular place or time and not in an easily identifiable manner. No one seemed to be doing anything clear and distinct that could be called accountability, or to be using any particular technology or standard for achieving it. This posed a simultaneously empirical and conceptual question: Where and when was the infrastructure of accountability, and how could it be described?

Our presentation for the workshop attempted to take seriously the challenges and frustrations that staffers experienced in relation to M&E. Describing M&E as elements in an information infrastructure under pressure, we aimed to stimulate a discussion of the difference between official assumptions about M&E and the composition of actual accountability practices. Specifically, we imagined that it would be useful for participants to consider how to prioritize the various competing goals embodied in their M&E work. But the ambition to create concrete recommendations for the application was quickly dashed. Somewhat surprisingly, the agenda for improved monitoring and evaluation was pushed from "below" rather than by management, who maintained that "We are doing well already."

What *we* got out of the workshop were the participants' reflections on the multivalent nature of monitoring. We ended with a somewhat incongruent list of important things to bear in mind. On the one hand, M&E was seen as a formal set of activities that "make up only a tiny part of what we do," but on the other it was also described as crucial for the creation of a knowledge base about the progress of projects. It was seen both as a way of acquiring information about "how to best use our and the donors' money" and as an entry point for participation that could be used to "get to know our target groups better." Finally, our participants thought that M&E had "huge potentials," although what these potentials were varied, and how they might be drawn out remained unclear.

In short, the workshop made it clear to us that the M&E infrastructure was characterized by multiplicity—that it contained several co-existing and possibly contradictory purposes and concerns (Mol 2002), which were therefore hard to manage. (See also Winthereik 2010.)

The Midterm Review

So far we have provided an extended "background" or "context" for NatureAid's infrastructure of accountability. The perspective we had gained from our location at the Danish office showed that this infrastructure was far from simple. It comprised an NGO operating in an environment characterized by competing agendas for conservation and socioeconomic development, changing funding modalities, and a plethora of formal and informal practices all of which were used to do M&E. Since these elements were related in uncertain ways, both in our view and in the minds of our informants, we wondered how they cohered, and indeed whether they hung together at all.

Luckily, while grappling with this question, we were offered the opportunity to join a review visit to the Vietnamese eco-tourism project. In retrospect, it became clear that a perspective from our informants' field sites was missing. Specifically, this field trip made explicit a concern with which

NatureAid's professionals constantly had to grapple: "how to handle local divergence from a desired norm; including how (or whether) to keep users from working around or outside the system" (Edwards et al. 2009, 366).

Magnus' way of handling this question, reiterated on several occasions, was to refrain from overemphasizing formal accountability and view monitoring "in the broadest possible sense." Only by leaving the Copenhagen offices in order to follow the infrastructure to some other places (where it would still, in some sense, have to hold), and then returning, did we become capable of appreciating the import of this injunction. For practical reasons, only Brit joined Magnus and Eric for the midterm review. In Vietnam, the members of the Danish team joined up with their Vietnamese partners: Nhi, a Hanoi-based project manager, three on-site project workers, and an on-site project manager.

The eco-tourism project had a somewhat unusual history. As part of a previous collaboration that had aimed to create a basis for sustainable agricultural production on the borders of a national park in Southern Vietnam, Magnus and Eric had been approached by the park's management. They had been told that the park had no way to prevent private-sector businesses from building hotels in the park under the rubric of "eco-tourism." Such building activities posed a significant threat to local wildlife. Both the park's management and the Danes were alarmed by businesses' promiscuous use of the term, which simply meant "tourism in nature" rather than "in cities." Thus, Magnus noted, "it was high time that we got an idea what eco-tourism actually meant to them." For Magnus, however, this "them" implied not only businesses, but also the park's management, local communities, and government agencies, since everyone had different interpretations of what "eco-tourism" implied.

The confusion was exacerbated because no policy or guidelines for eco-tourism were available. In this situation, Magnus and Eric envisioned a small project that could work as a showcase for eco-tourism and, so they hoped, might be used to influence future initiatives and policies in the region. Consequently, they arranged a meeting with the vice-director of the park. That meeting led to a workshop with representatives of the park's management, villages, regional government offices, and private-sector travel agencies. The workshop was a huge success, Magnus explained, because of the participation of multiple stakeholders.

Accordingly, NatureAid went forward with a project application in collaboration with Vietnamese partners and received funding. From being the instigators of the project, Magnus and Eric gradually switched roles to their position at the time of our visit, so that their main task was to ensure that the Vietnamese counterpart ran the project according to the "immediate objectives" of the project proposal. Meanwhile, the Danish office continued to hold responsibility for delivering technical assistance on capacity building and stakeholder involvement.

By reading through various reports produced over the past years (a baseline study and five evaluation reports), one can learn about the successes and disappointments of this project. (Cf. Mosse 2005b.) Some of the reports are surprisingly candid. As the project matured, they begin to resemble internal working documents—complex to-do lists—rather than mechanisms of formal control. The first report exhibits the excitement of the initial, successful workshops. Its main message is that the project has the potential to become a showcase for eco-tourism in Vietnam. Yet the second report tells about practicalities that hinder progress: difficulties in obtaining the right permissions and finding qualified staffers. Still, hopes remain high. The third report excitedly announces that a project coordinator has been hired. But the fourth report notes that "implementation is extremely delayed." It is full of action points. The fifth report tells about the new Ecotourism Management Board that has been set up at the district level, and about the election of a Village Management Board. The election, in which pebbles are used for voting, is clearly a courtesy to the multi-stakeholder ambition of the project.

All of the reports are co-authored. The normal procedure is that the Danes draft them on the basis of input from their Vietnamese partners, who then comment on the drafts. As we shall see, the partnership relation that facilitates this interactive drafting process is crucial for the way a project can be known.

Organizing Collaboration

Collaboration and participation are at the heart of both the project design and of the midterm review. The participatory ambition of the project is most clearly stated in the first "immediate objective":

> Ecotourism planning in and around the . . . National Park is carried out through a participatory multi-stakeholder process and is incorporated into development plans at commune, district and provincial levels

This, then, is an overarching strategic aim. The project goal reflects NatureAid's ambition to live up to its role as an NGO—that is, as an agent of civil society that facilitates a democratic process.[7] Nevertheless, implementing these objectives requires that many mundane problems be solved.

One impediment to progress of the eco-tourism project was that Vietnamese NGO professionals found it unattractive to live and work in the park. The review team needed to address this issue even though, in principle, it wasn't their responsibility. Yet because staffing was crucial, the review team had to get involved. On this occasion, as on many others, supervision and evaluation blurred.

While Magnus was planning the review trip, one of his main concerns was how to address staffing issues and other practical details without being overwhelmed by them. Handling such difficulties with a light but firm touch was important because the "real goal" of the visit was to document and assess the impact of the project. For Magnus, holding this balance entailed tricky questions such as "How to make sure the project is moving in the right direction without being too narrowly focused on what was promised?" and "How to be a good partner, while continuing to insist that certain issues must be taken care of?" Specifically, it entailed navigating the partnership in a way that would make it practicable to acquire knowledge about the project even while the question of what counts as relevant knowledge remained in dispute. It also required the Danish team to play the role of good partners in a way that could be appreciated and, ideally, emulated by other partners, even those whose views on partnership and M&E differed significantly from team's perspective.

To handle the double challenge of collaborating (enacting partnership) with a view to coming to know the project (monitoring the partnership), the Danish team decided that the midterm review should be organized as a series of meetings in the field—meetings that would include stakeholders in and around the national park where the eco-tourism project was implemented. This proposal caused immediate trouble because the project manager from NatureAid Vietnam thought the review could and should be undertaken as a desk study. In an effort to smooth over this disagreement, shaping the field trip even before it had begun, different versions of the terms of reference for the midterm review circulated. No common ground was found, however, and the Danish team insisted on going into the field. In spite of complaints, the Vietnamese project manager and her colleagues were thus obliged to organize the many meetings with stakeholders that the Danes requested. At the time of the arrival of the Danish team, the densely packed program had already been rescheduled several times; it continued to be modified throughout the visit.

Discussions about scheduling details offer something like a microcosm in which different commitments to accountability and partnership can be ascertained. Negotiations about who are the important collaborators and partners take place, in part, through the rescheduling of meetings and the organizing of practical matters. Just as the partners disagreed on the need for meetings, they also failed to agree on who were the important stakeholders and on what were the major issues to be discussed. At the heart of these tensions lay different understandings of the central question: What do we need to know for the review, and how can we get to know it?

First and foremost, Magnus and Eric worried about not getting to meet with all the relevant stakeholders. If they were unable to hold the right meetings, they would also be unable to gain knowledge about the project. Without such knowledge, they would be unable to help their partners learn, and thus the participatory approach would fail. Back home, a lack of relevant knowledge might jeopardize their ability to obtain funding for eco-tourism projects in the future. Thus, from the point of view of the Danes, meeting all the involved people was a non-negotiable requirement. However, from Nhi's perspective a desk study was quite appropriate. After all, the formally stated purpose of the midterm review was precisely to "weigh and measure" the project's success. Indeed, Nhi saw the field visit as pointless or worse. It threatened simultaneously to be inconvenient and to bring to light complexities of the project that would have to be handled. Thus, the partnership was in tension from the outset of the review visit.

Negotiating Participation

Although the Danish team would eventually need indicators for accountability purposes back home, numbers are far from their preferred form of knowledge making. Indeed, they viewed indicators as void of context and thus crude. But if indicators are not sufficient to establish whether the "immediate objectives" are being met, what alternative tools might be used? In general, the emphasis of the Danish team is on making knowledge through participatory processes.

Days are filled with meetings, from early morning to late evening. Indeed, meetings are the most common setting and technique for making project knowledge. There are internal meetings with NGO staffers, meetings with government officials, meetings with private investors, meetings with villagers, and meetings with local authorities. Some take the format of friendly conversations; others are question-and-answer sessions. Some take place around a table or next to a whiteboard; in others we are seated at the floor. All of the meetings involve drinking green tea.

Almost ritually, all meetings also involve an opening speech in which Magnus lays out the general purpose of the review and the specific aim of the meeting. The opening comes in a number of variations, but it always carries the same message: "We are here to learn about what has happened in the project so far, but we also want to look forward."

In contrast, no meetings begin with a discussion of outcome measures. Instead, the Danish speakers consistently emphasize the importance of qualitative change: Has the project initiated a "positive development" with respect to sustainable and community-based tourism? Has it been capable of "raising awareness" about the importance of protecting the natural environment among the shareholders? Even apparently quantitative questions, such as whether there is evidence that the project has lessened pressure on the forest and created a sustainable source of income, are seen as too complex for indicators. The uncertainties of quantification are perceived by the Danes as simply too great in the context of the still-emerging eco-tourism project. Consequently, indicators are viewed as inferior to participatory, qualitative forms of knowledge making. As far as the Danes are concerned, indicators are decidedly *not* the right tools for the job (Clarke and Fujimura 1992). Nevertheless, the midterm review will eventually have to be shaped around them.

At the same time, Vietnamese partners joke about "the participatory approach." During a meal, Magnus inquired about the constitution of a management board in one of the villages. There were two women and two men on the board. Nhi, the Hanoi-based local project manager, made a dismissive comment: "'Gender balance,' you know." On another occasion, she took Magnus to task: "Did he forget," she asks, pronouncing each syllable individually, "that they are supposed to adopt a *par-ti-ci-pa-to-ry* approach?"

Of course, these situations exhibit the fluency of Vietnamese staffers in development discourse. They use it strategically and practically to communicate with Magnus and Eric, and they probably also use it among themselves. This comfortable ownership offers a clue to why analyses that focus exclusively on the "tyranny" of participation (Cooke and Kothari 2001) miss something important. If partnership is tyrannical, its effects are also felt by Magnus and Eric. To some extent *they* come across as "oppressed" in these situations. After all, participation is specifically *their problem*: it is part of their job description to make participation materialize during meetings, though few others seem interested.

Even so, it is striking that the words *control, monitoring,* and *evaluation* are so systematically not used in conversations and meetings with local managers. Is the NGO's staff hiding its control ambition under a veil of *pro forma* participation? Asked directly, Magnus brushed off this suggestion with the following observation: "They know *very well* that we are also here to monitor and control." The notion that monitoring and evaluation are forms of external control that are hidden behind a veil of partnership is far from the interpretation of the Danish team members.

In fact, they see the situation almost in reverse. In their view, the problem is that the Vietnamese are so accustomed to control, and expect it to such an extent, that they fail to realize that shared learning is *really* crucial to the Danes' participatory approach. Magnus and Eric continuously emphasize learning and participation *because of* these discrepant interpretations of what partnership entails. In spite of these efforts, however, both their partners and government representatives seem more than a bit confused about the idea that the project can be assessed based on loose conversations about "what goes well."

If we emphasized the opening ritual of meetings, it was thus to point to the double role of the review team. Precisely because the team is only too obviously there to monitor and evaluate, the meeting becomes a crucial technique for the enactment of partnership and participation. Thus, Magnus and Eric aim to offer to their Vietnamese partners concrete illustration of how it is possible to press agendas from within a participatory framework. Such exemplary partnership behavior is exhibited, for example, in briefing meetings with the local project manager during which the team carefully refrains from interfering so that she can take charge of the process. As part of its learning objective, then, the team also seeks to implement ideas about what it means to be a "good partner." But good partnership is itself a term in friction.

Moments of Partnership

The field visit involves an overnight stay on a hill near the border of the national park, where the project has provided support for the building of a traditional long house. The long house will function as a base camp for tourists who want to experience the wilds of the forest and the culture of the local villagers. A private tour operator will "supply" tourists to the house, which will be staffed and managed by the villagers.

A trip has been planned so that the reviewers can act as tourists and try out some of the services and products that the eco-tourism project will be offering in the future. This seemingly innocuous activity brings to light different views about what counts as relevant knowledge about the partnership.

The Danish members look forward to entering the forest, guided by villagers; they are eager to experience this concrete outcome of the project. Joining the trek feels like a natural and fun thing to do. Further, this "embodied participation" enables them to generate knowledge that is relevant for monitoring and evaluating the project. Taking on the role of faux tourists, they are enabled to see how villagers take on the role of tourist forest guides.

In contrast, the Vietnamese project manager, Nhi, is quite reluctant to join the field trip. From early on, she complains about forest inconveniences and is adamant that she isn't going on the hike. As she sees it, there is no way in which hiking is a relevant part of the collaborative context: it neither enacts participatory ideals nor produces relevant knowledge. Instead, she is concerned with an altogether different series of troublesome issues, none of which have to do with her professional identity or aims: hikes through rugged terrain, mosquitoes, noises of the forest that keep her awake at night, and ghosts. More than simply exhibiting recalcitrance, her concerns point to the particularity of the expectations and interests that undergirds the notion of tourism in the "wild." They allow us to recognize the distinctly European-American flavor of eco-tourism, an issue that is also brought up later in conversations among the partners.

Although the hike is meant as an exercise in partnership, then, it is also indicative of friction, if not breakdown. While creating closer relations between the Danish team, the villagers, and those responsible for the training of cooks, hosts, and forest guides, it also intensifies disagreement with partners from NatureAid Vietnam. What partnership entails is itself at stake in the decision whether or not to go on the hike.

It therefore comes as a pleasant surprise to the Danish team that Thanh, the project manager based in the national park, decides to come along. Thanh, who has a past as a civil servant, sets out on the trek wearing a spotless white shirt, as if going to the office. He returns with torn clothes and ants crawling on his back; to recover from the fast-paced walking, he had taken a rest on the forest floor. His change in attire and comportment almost literally embodies the tension between the modes of learning advocated by the members of the Danish team and their Vietnamese partners.

As a moment of partnership, the hike in the forest gives expression both to collaboration and to its frictions.

The Long House Meeting

Aside from the trek, the days in the field are spent mainly in meetings. Representatives from the management boards of two villages, the management board of the long house, the Women's Union, and the Youth Union all visit the long house. On the night before the first meeting, many villagers stayed late as part of a cultural performance—one of the services, along with cooking and forest guidance, that the villagers have been trained to offer to tourists. Perhaps in consequence, few people arrive for the meeting on the next morning; those that do show up seem tired. With regret, Magnus and Eric note that some villagers seem to represent both the Women's Union and Youth Union and the management boards.

All are seated on the floor of the long house. As usual, Magnus starts the meeting by explaining its purpose: to "look back" on what has taken place, as well as to "look forward."

Trang, the project's accountant, translates. Nhi leans back against the wall with an exhausted look on her face. The conversation goes slowly. It seems as if everyone has come to get something, and that thing isn't materializing. The Danish team members are quite explicit that they want to hear about the villagers' experiences with building the long house and about their hopes for the project. They hope the villagers will have suggestions or ideas for how to improve the project's activities. But the villagers don't volunteer much information. It seems that they, too, are looking for something that is absent. They have spent time and resources building the long house and have carved a totem pole of a kind that normally is made only for the buffalo festival. Making the pole was expensive, but they made it for the project as a joint investment in the future. So when will the tourists arrive?

Despite the sullen atmosphere, after the meeting Eric exclaims "This is the stuff that makes you high!" In a way, it isn't hard to understand his need to smooth over the embarrassment just experienced. The meeting clearly demonstrates some less exciting aspects of making knowledge through partnership. After all, it isn't particularly great to travel the world to sit on a wooden floor and experience an information deadlock. It isn't particularly exciting to find that one doesn't get much help from one's collaborators, who roll their eyes and fail to translate.

After the meeting, Magnus and Eric are therefore frustrated. For one thing, they aren't sure that the right people participated in the meeting. But mainly they feel let down by their Vietnamese partners. Without their Vietnamese partners' active support, Magnus and Eric say, they instantaneously turn into "big, white, clumsy men." They wanted to see Nhi, Thanh, and Trang play active roles. They had hoped for an exciting exchange about shared goals, practices, and improved methods. Their frustrations indicate how hard it is to be a good partner with people who see partnership differently. But if one thing the meeting highlighted was that the participants have different understandings of what partnerships entails, then how should this experience, this knowledge, be transported back through formal infrastructures of accountability?

Figure 5.1
The gallery of the long house.

Opening Indicators

A meeting in the main building of the National Park is about to commence. This time it is an internal meeting for the Danish team and their Vietnamese NGO partners. We are past opening rituals: this meeting features neither a celebration of progress nor invitations to participate. The discussion is finally turning to the question of how to specify the right indicators for the midterm review. It is late in the afternoon, and the atmosphere is again tired. Everyone fiddles with his or her own laptop computer. Magnus urges the Vietnamese participants to talk about the project. He wants to know what is going well, what should be adjusted, and which initiatives are not succeeding. His attempts fail to arouse much interest. Giving up on his participatory strategy, Magnus suggests that they walk slowly through the indicators to get an overview.

Once again, Magnus is trying to be engaging and to involve everyone. He keeps inviting everyone to join a dialogue, but if the dialogue is to succeed then someone will have to take up the offer. Experiences like this team meeting gives life to a theory of collaboration that Magnus aired while on the plane: "It's like when Bourdieu talks about habitus. Each party is present with his or her habitus that cannot easily be changed. What can be changed is that you can make more transparent where you are coming from." Considering each participant as embodying a particular habitus that is basically unchangeable makes perfect sense during the team meeting. The interaction seems blocked; there doesn't even seem to be friction, only detachment.

In this case, however, something unexpected happens. Thanh puts his laptop down, stands up, and writes some numbers on the whiteboard. They represent the park's investments in eco-tourism activities. He explains that he has received the figures from the park's management. This is a leap forward because the park's management has so far not been very open with its budgets. Thanh certainly seems to be quite satisfied to have brought something this specific to the table. His Vietnamese colleagues are less impressed: the figures cover a ten-year period and are not very precise. Still, Nhi gets up to make some calculations on top of Thanh's scribbles. Suddenly everyone is involved in a discussion about park investments and strategies and how they relate to eco-tourism in the region and nationally.

After the discussion, Magnus continues the walk-through of the indicators. The atmosphere changes and, as if magically, the discussion has turned lively and exciting. It has also expanded. It is now simultaneously about the nature of indicators as such (What kind of tool is an indicator? How flexible is it?) and the worth of statistics (How is poverty measured in a Vietnamese context? Can the statistics be trusted?). It is also an occasion to discuss the sensitive fact that only one out of the three communities that participate in the project has had much success so far.

One of the indicators put on the table is that 30 percent of the tourists in the National Park visit the villages, and the villages benefit from this. Nhi argues that 30 percent is a very high number. Now the park has 20,000 visitors per year. But the vast majority of these visitors are Vietnamese, and very few Vietnamese are interested in going to the villages. (Perhaps the potential pleasures of eco-tourism do not appeal to their habitus.) Would it be more sensible to specify that 30 percent of the *international* tourists visit the villages, and that the villages benefit from these visits? Then a new question is raised: How to monitor whether the villages benefit? It is very difficult to measure how many tourists actually visit the villages, since the park has neither baseline figures nor an operational system for documenting visitors. The discussion turns the self-explanatory indicator into a figure of uncertainty. The group agrees to think about what would be a more realistic goal to inscribe in the midterm review report.

Another indicator is scrutinized. It states that "100 households will no longer be categorized as poor by the end of the project." Trang says that in the involved communities 120 households in total are categorized as poor. The indicator seems to suggest that the project will more or less eradicate poverty in the villages. "This is unrealistic," she says, and people nod. When Eric asks more questions about the figures, though, Trang produces a document that shows the project to be quite successful. People gather around her laptop. In one village, 67 percent of households were categorized as poor in 2009, but in 2010 the figure was much lower, only 30 percent. This is great, everyone agrees, but it is also quite strange, since the project has not yet produced any income opportunities. So how can the poverty level be so much lower? Gradually it dawns on the team that the reason is that the government has changed the poverty baseline. Nhi interprets:

> It is quite simple. Everybody wants to get out of poverty, so it is a good thing if the government can produce figures that show a reduced poverty rate. At the same time, the poor get government benefits, so sometimes the poverty line is defined by how much money is in the treasury. Sometimes the province publishes funny statistics that do not add up to 100 percent. The numbers change all the time.

At this point, Magnus suggests that they work on further specification of the indicators by email, and urges Nhi to think about appropriate definitions. Obviously, he is concerned about maintaining shared responsibility for adjusting the indicators after the meeting is over. Although the conversation flows during the meeting, the moment of partnership may well have disappeared by the time the team is back in Denmark. At this point, the problem of finalizing the midterm report will be squarely back in the hands of the Danes.

Recontextualizations

Marilyn Strathern (2006) has argued that technologies of accountability simultaneously make social relations, create forms of organizational visibility, *and* constructs new forms of opacity. The midterm review, too, links the visions and concerns of Danida with those of NatureAid Denmark, the two organizations' Vietnamese partners, and the villagers. It connects the project's original design—the project on paper—with the project as it unfolds concretely in a Vietnamese park.

Clearly the indifferences, the confusions, and the occasional pleasures that make up moments of partnership and shared knowledge making cannot easily be articulated in a format appropriate to formal reporting. As techniques for knowing, standardized representational formats demand the deletion of knowledge about their own messy conditions of existence. These formats oblige their users to propagate objective-sounding ideas about monitoring and evaluation—ideas far removed from actual knowledge-making processes. This is the built-in paradox of practical aid ontologies that rely on "global" infrastructure to make "local knowledge" visible. It is obvious that a report can have only a limited set of contact points with the collaborative encounters, dense with affect and friction, that we saw unfolding during the field trip.

None of this is very surprising. After all, formal reports are precisely where we would expect decontextualization to occur. Local practices and collaborative tensions, we are often told, necessarily vanish when translated into formal systems of representation (Li 2007; Peluso 1991; Shore and Wright 1997). Yet it is because team members regret this situation—perhaps as much as critical ethnographers might—that they struggle to make these contact points explicit in the reports they bring back to Copenhagen.[8] The surprise, then, is that some of the collaborative frictions do make their way into the midterm review. An example is found under the heading "Participatory approaches":

> The Vietnamese project staff is well aware of the importance of applying participatory approaches, and even makes jokes about it, as this has been repeated over and over again by the DK team. There is no doubt that the approaches of the project staff have become more participatory during the last months, but still there needs to be a more stringent focus. (Final Review Report of April 2011)

The style of writing is positive, even self-depreciating. Acknowledging that "for both parties, the project has been exciting, innovative, challenging and hard work," misunderstandings are not weeded out.

On the one hand, then, the authoring team tinkers with reporting language in order to maintain a sense of the complexity of partnership and collaboration (even as these uncertainties are given expression in the form of supposedly manageable action points). On the other hand, responsibility for the problems encountered in the partnership is also placed in the hands of Vietnamese partners. And why? Because of the problems Vietnamese partners have understanding what partnership *really* entails.

Thus, the paragraph quoted above ends by emphasizing Thanh's difficulties in adjusting to the collaborative efforts embedded in the project: "The project coordinator has a government background and is used to the top-down system and has had severe difficulties getting accustomed to applying participatory approaches." The section leads to action point 39, which suggests that "the project team should always apply a facilitating approach and make sure to include the target groups in all parts of the development steps."

Monitoring "in the wild," we have suggested, happens through moments of partnership whereby the project comes to life at sites of collaborative friction. Weedy infrastructure emerges in the interstitial spaces where different partnerships knowledges remain in tension. Yet these moments remain unconnected; they do not easily align with the formal infrastructure through which knowledge travels back to the donors.

Thus, in spite of the team's efforts, it is clearly not the case that all relevant complexities of accountability and knowledge making are retained in the midterm review. Indeed, it is hard to imagine how it would be possible to keep moments of partnership in formal view. Yet it is also hard to figure out precisely what would be the purpose of doing so. After all, we are witness to a process of infrastructural recontextualization rather than decontextualization. The final report *must* enact a project world that speaks to the institutional requirements and interpretive demands of donors.

Thus it is crucial to bear in mind that the midterm review has other purposes than enacting complexities encountered in the field (Harper 1998). Indeed, its purpose is eventually to perform the reality of the eco-tourism project in a manner suited for the institutional environments of funders. Just as partnership must be handled with care at meetings in the long house, so it must be cared for amidst the bureaucracy of Danish donors.[9] The problem is that the required forms of care differ. Yet even though the practical aid ontologies enacted in these settings are radically divergent, they still must be made compatible, at least minimally, through formal infrastructure. Otherwise there will be no development partnership at all.

In this situation, it is necessary to use the distinction between M&E in formal paper worlds as opposed to "in the wild" with finesse. From the point of view of Vietnamese villagers, the "wild" of development is located in the exotic settings of Danish aid organizations. Indeed, the mutual exoticism (see Viveiros de Castro 2004) of the partners accounts for many of the troubled interactions that occurred during the supervision visit.

As we have shown, it was not only to us, as observers, that the shifting contexts of partnership posed problems of comprehension and compatibility. Rather, shifting contexts, agendas, partnerships, and purposes seemed to form an unstable ground for all the partners. The problem faced by all partners was that what each of them saw as relevant context, at any moment, could not be assumed to be shared. In such a situation, what seems somewhat astonishing is that, most of the time, the infrastructure of accountability actually holds together well enough to keep the project on track.

A Question of Cultivation?

In this chapter we have followed the work done by the conservation team to facilitate monitoring and evaluation of their projects and accountability in relation to both partners and funders. Throughout the chapter we have played with two dualisms that are tempting to take for granted when trying to make sense of the case. The first dualism opposes the "formal world" of official reports to the "wild" of a project's realities. To an extent this dualism helps us to grasp the fragility of this infrastructure of accountability: it helps us to see that it is required to live up to almost incompatible demands. Stretching just-about-incommensurable worlds, the infrastructure wears thin. The second dualism contrasts the professionalized world of Danish NGO employees with the lived worlds of Vietnamese aid recipients. This enables a critical approach to partnership, seen as an ideological smokescreen under which the power of development unfolds.

Though the pictures painted by these dualisms are not completely false, they should also not be taken too seriously. Attentiveness to infrastructures of accountability shows that relations between "the formal" and "the informal"—that is, between "the professional outsider" and "the villager"—take multiple forms. What are the implications of this observation? Our answer contrasts with an analytical notion proposed both in the anthropology of development and in the design of information systems. What we have in mind is the idea of *cultivation*.

Why would one talk about development and infrastructure practices as a matter of cultivation? David Mosse's book *Cultivating Development* (2005b) offers a cogent answer. Mosse wants to find a middle ground between instrumentalist understandings and critical social and cultural anthropology. He analyzes development practices from the inside of an Indian aid project. The point is to create more nuanced understandings of the complex realities of projects, and to find ways of improving them. Cultivation happens through gradual processes of relation making that result in mutual trust and learning among aid partners. Mosse concludes that "the emancipatory intentions of policy and the aspirations and interests of the poor *can* be linked" (240) even if the circumstances are difficult.

Interestingly, cultivation has also been proposed within the context of information-systems research, as a metaphor that might replace the idea of rational design. Thus, it has been argued that infrastructures can be seen "as a sort of a living organism that can be cultivated, rather than some dead material to be designed" (Hanseth n.d.).

Cultivation is such an appealing idea because it suggests gradual evolution and growth, not just in size but also in the richness and quality of relations. Further, cultivation is likely to be sustainable only if cultivators are attentive—if they care about that which they hope to "grow." Thinking along those lines might lead us to imagine collaborative challenges as requiring the cultivation of better communication practices and infrastructures. On the side of development organizations, for example, reporting formats and standards could be made more flexible and more open-ended. And in the villages, even more effort could be put into cultivating an atmosphere congenial to explication of people's real opinions. Perhaps the result would resemble a Habermasian "ideal speech situation" more closely than a meeting of unchangeable Bourdieuan "habituses." Weaving these threads together might facilitate the cultivation of a more resilient infrastructure of accountability.

Even so, the notion of cultivation is problematic. For one thing, it is based on an image of orderliness, or even a sense of environmental control. (See Drayton 2000.) As Bauman (1990) reminded us, it is, after all, a *gardener* who cultivates; plants *are cultivated*. The term *cultivation* distributes agency asymmetrically. It is telling, and more than a bit curious, that cultivation has been proposed as a conceptual tool in development anthropology and information-systems research, both disciplines that deal with heterogeneous sets of recalcitrant others.

Weedy Infrastructure

Nevertheless, we remain quite taken with the analytical possibilities enabled by using organic imagery to think of social relations.[10] The anthropologist Anna Tsing offers a different image. She induces us to think carefully about weediness: "For an observer used to imagining agriculture as cleanly weeded lines of corn, wheat, or tomatoes, but not all tossed together, the most amazing thing about a . . . swidden field is the extraordinary number of plants growing together in the same small spot" (2005, 165). Still, Tsing continues, a swidden is likely to be perceived by "visitors from the city" as "'natural' forest" (174). Although most of the variety found in swiddens is of a "*cultivated* variety," visitors arriving in these fields cannot see that this is an already ordered space. They notice only wilderness.

The reason swiddens seem wild is that they are organized according to principles that a visitor can't recognize. This inability to recognize existing order is consequential. If swiddens seem wild to newcomers, they will have no qualms about clearing the land to cultivate their own crops. The landscape becomes a palimpsest as one kind of cultivation is laid on top of another. No wonder confusion and controversy emerge.

Infrastructure and partnership can be characterized by the way in which, palimpsest-like, they grow on top of each other, producing weeds inside one another. In Denmark, team members were engaged in ongoing attempts to cultivate new, more flexible M&E systems. They were willing to test our expertise precisely because they didn't know how to proceed. Far from resistant to formalization, these professionals *would like* tools to help them order—cultivate—their informational practices. Yet, thanks to what they already knew about the limits of formal M&E, they also sensed that formal infrastructure was an unlikely candidate for managing weedy partnerships.

And not without reason. For as we have seen, what counts as cultivated and what counts as weedy differs for donors, team members, Vietnamese partners, and villagers. Participatory meetings and hikes, which Danish team members perceive as central cultivating principles, are perceived as weeds by Vietnamese partners. The formal, bureaucratic, "non-collaborative" behavior of Vietnamese partners is a weed from the point of view of participatory methods and partnerships.

This is where the notion of cultivation might be brought in, but not in a positive sense. The midterm review's characterization of Thanh's "severe difficulties getting accustomed to applying participatory approaches" makes it plain that the Danish team thought Thanh needed cultivation. In this context, then, cultivation points to asymmetrical agency in the partnership, not harmonious agreement. Cultivation is required just at the moments when "weeds" become intolerable and methods must be applied to "weed them out."

But of course weeds are also part of the rationale for developing formal and technical infrastructure in the first place. Without the mutual misunderstandings, hopes, and aspirations that make up project relationships, there would be no project, no rationale, no partnership, and thus no knowledge and nothing to report. The infrastructural challenge arises from the tension between the "installed base" of formats and requirements, which the organization needs, and the weediness of partnership relations, which it also needs. If NatureAid Denmark's infrastructure of accountability is fragile, an important reason for the fragility thus has do with weeds that constantly threaten to emerge from the cracks between formal systems. But noticing only the *threat* of weeds, and dreaming of cultivation, is an inadequate response. In particular, it keeps people from recognizing how, in moments of partnership, weeds make infrastructure work.

In her discussion of swiddens, Anna Tsing notes that "weeds have been of little interest to conservationists; we think of them only as indicators of disturbance" (2005, 174). The observation is strangely resonant because our informants are members of a "conservation team." Even so, as we have shown, they don't view *all* weeds as "indicators of disturbance."

There and Back Again: An Inverse Development Anthropology

Our story began, in Copenhagen, with the question of how to identify and characterize the infrastructure of accountability through which NatureAid Denmark knew its projects and made them knowable to other stakeholders, including the Network and funders. Eventually we were challenged by our informants to interpret this infrastructure "in the broadest possible sense." Following this advice, we found that what counted as relevant context in this partnership had to be continually performed in a variety of collaborative, negotiated encounters. (See Verran 2002.) In different ways, all of these encounters related to the creation of formal reports that would be capable of moving through the infrastructure of accountability. They also highlighted that development partnerships involve navigating divergent perspectives on what infrastructure and accountability are.

The efforts and understandings of Danish team members have obviously been central to this analysis, which has not focused equally on the perceptions and interpretations of the partners working for NatureAid Vietnam or the villagers involved in eco-tourism experiments. Even so, our ambition is not to rectify the perspective of the Danish team.

Critical development anthropology has been harshly critical of development initiatives and projects. Drawing on Marxist and Foucauldian perspectives, many studies, including classical ones by Arturo Escobar (1995) and James Ferguson (1990), have criticized development projects for their Western-centric, paternalistic, or imperialistic ideologies. They have often done so from an ethnographic location among villagers. From such a location, it has often been practicable to document that foreign consultants entered local sites with plans and projects that made little sense, or were doomed to fail, but were nevertheless carried out because development organizations knew little about, or were indifferent to, context.

In contrast, we shared ethnographic location with the aid professionals. Our story of aid infrastructure began in an office in Copenhagen, our "real" field site, from whence we traveled with our informants—development consultants—to *their* "field site." Going "there and back again" has enabled us to tell a quite different story. We think of this analysis as a case of *inverse development anthropology*, since, compared with critical anthropology of development, the sequence and the direction of the ethnography are reversed. But the point is not *analytical* reversal.

Critical development anthropology tends to enact development workers as Others. In many accounts, development workers are presented as ignorant, patronizing, and very powerful all at once. Indeed, Ferguson describes such morality tales as an "anthropological specialty" (1997, 165). In contrast, we have traced infrastructures of partnership and accountability and found them to inherently fragile. The development partners need each other, just as they need the formal infrastructure—which is nevertheless bound to fail to represent the complexities of interactions. Weediness is at once constitutive and inhibitive of aid infrastructure. It is at once problem, solution, and fact.

The endpoint of this chapter was the completion of the midterm review report, but the report is still moving. For one thing, it travels to the Ministry of Foreign Affairs, where it is read and evaluated by funders. Is this then the endpoint of auditing? No, because the practices of the ministry are also monitored and evaluated. Development workers regularly mention that, ultimately, they are accountable to the taxpayers. In the Danish context, this final form of accountability is ensured by the audits and evaluations carried out by the National Audit Office. In the next chapter, we therefore follow monitoring one final step upstream, in order to grapple with how the National Audit Office knows the realities of aid. We find, to our surprise, that the National Audit Office is *also* trying to carefully balance control and learning, formal rules, and informal participation.

6 Wormholes: Loops of Auditing and Learning

Auditing and Accountability for the "Taxpayer's Money"

How is the monitoring work carried out by the National Audit Office in central Copenhagen related to the monitoring work of NatureAid? Among other things, the civil servants, the environmentalists and the people living in and around threatened natural reserves are connected through a flow of documentation through which they seek to account for the progress of development projects. Through multiple procedures and formats, they seek, in their quite different ways, to understand the quality of aid projects and their effects. But there is one central difference. Whereas NGOs such as NatureAid are required to monitor the progress of its own projects, the task of the National Audit Office is to audit these organizations. The National Audit Office provides the final external check on how NGOs spend funds they have received from the Danish state.

Recall the argument (made by development consultants) that aid development partnerships are necessarily asymmetrical because donors are first and foremost accountable to their own taxpayers. They must be able to prove that tax money is correctly spent (that is, according to rules) on development projects, and evaluate whether it is spent well. If it isn't spent correctly, the organizations are breaking Danish laws. And if it isn't spent well, political and public support for aid expenditure may well dwindle, as it has threatened to do, in Denmark and elsewhere, in recent years.[1]

For good reasons, donors are not in charge of providing the ultimate evaluation of how well they are doing. That is the task of audit offices. The National Audit Office, therefore, is supposed to be the final link in the "chain of accountability" that guarantees the transparency of aid projects. This "chain" includes non-governmental organization and their partners, funders (such as Danida), and Danish taxpayers. However, as we shall see, auditing "loops" in such a way that to talk about an infrastructural endpoint to it becomes complicated. There really isn't any "chain" at all.

The National Audit Office is an organizational body of considerably importance and authority. Since it is the final auditing instance, it is incumbent that its audits are objectively correct. Auditing guarantees that public money isn't squandered. For the same reason, the legitimacy of the National Audit Office is entwined with its ability to make its auditing objective.

The National Audit Office's procedures, however, are themselves enacted in and as an infrastructure of loops between auditors and auditees. These loops put auditors in a position in which, while interacting with the organizations they audit, they must remain detached from them. In this chapter, we argue that the tension between detachment and interaction creates "wormholes" through which auditing passes, mutating as it does so.

The National Audit Office

Denmark's National Audit Office has undergone substantial organizational changes since 1986.[2] It has halved the number of office workers it employs and increased the number of employees with academic backgrounds, mostly in social science.

The mandate of the National Audit Office is to audit the state's accounts. More specifically, it is to verify that funds are used according to parliamentary decisions. Organizationally, the office reports to the Public Accounts Committee, yet it conducts its tasks independently to ensure impartiality and objectivity. In general terms, the prevalence of audits testifies to the strength of the *audit explosion* thesis (Power 1994). However, although the National Audit Office appears to be the epitome of external control, its modes of operation create organizational and epistemological frictions that, we suggest, give rise to *audit implosion.*

The more than 250 employees working in the office come from a wide range of backgrounds "reflecting the many tasks that the institution is performing."[3] These tasks include forms of auditing that go considerably beyond traditional evaluations of accounts and budgets (which of course remain central). "Auditing in many ways"[4] is the expression used to describe these activities on the office's homepage. Carrying out these diverse tasks involves addressing questions such as what should be audited and where, as well as questions about the use of methods and interpretation of data. The variety of critical decisions that must be made before and during an audit provided us with an initial sense that, although the office is committed to an ideal of procedural objectivity, this ideal co-exists with a far more mixed and diffuse sense of obligation to assist organizations and help them learn.

Yet this ambition may create tensions, since too much "assistance" may be seen by skeptical outsiders as demonstrating a lack of objective impartiality.

Accordingly, the auditors with whom we spoke are in a situation that is both practically challenging and epistemologically interesting. They deal reflexively with questions about validity, objectivity, and methods that are reminiscent of issues that continue to haunt social science more broadly. However, since the office operates with a political mandate that relies largely on a positivist and quantitative understanding of what performance auditing is and what it can and should do, they navigate difficult terrain.

Audit Explosion

According to Michael Power's famous diagnosis, European-American society became an audit society around 1994. The word *audit* was used with "growing frequency" (Power 1994, 6),[5] and audits were conducted in more and more contexts. "In addition to financial audits," Power wrote, "there are now environmental audits, value for money audits, management audits, forensic audits, data audits, intellectual property audits, medical audits, teaching audits, technology audits, stress audits, democracy audits and many others besides." (ibid.) Indeed, Power suggested, an audit explosion could be witnessed in the mid 1990s.

As an organizational and societal tool, auditing is intimately bound up with issues relating to control, accountability, and transparency. Auditing is not only or even primarily concerned with the measurement of performance; instead it concerns evaluating the governing qualities of systems. It is about the "policing of policing" (Power 1994, 11). Power's analysis emphasized that this meta-governance relied on quantification to an unprecedented degree. His critical observation was that the excessive interest in quantifying practice risked undermining trust, dialogue, and autonomy and replacing these values with measurement, expertise, and discipline.

Power's analysis was picked up by scholars in various fields, including organizational theory (e.g., Munro and Mouritsen 1996) and social anthropology (e.g., Gefou-Medianou 2000; Pels 2000; Strathern 2000a). Anthropological commentators, in particular, have argued that although the audit explosion undoubtedly has engendered organizational change processes, another pertinent issue concerns the implications of auditing understood as cultural form.

Seen as cultural form, auditing becomes amenable to ethnographic scrutiny. As Marilyn Strathern suggests, such scrutiny would facilitate the delineation of "some of the scale and pervasiveness of the way in which the twinned precepts of economic efficiency and good practice are being pursued" (2000b, 2). Such an approach would entail a simultaneous investigation of the conventions of audit modes of operation and aspiration, and a diagnostic exploration of how auditing may be morphing into new social and cultural shapes. Exploring the ubiquitous practices of monitoring that unfold in the Danish National Auditing Office's department of performance auditing, the present chapter is guided by these lines of inquiry. Of course, in one sense morphing is inherent to any practical ontology. It is also not new to auditing, which in the 1990s was radically transformed by its novel engagement with "ethics" (Strathern 2000a). Yet, we argue, audit implosion designates a very particular form of ontological shape shifting.

From early on, Michael Power himself asked what is "beyond audit" (1994, 32). In line with his general critical argument, he suggested that "we seem to have lost an ability to be publicly skeptical about the fashion for audit and quality assurance; they appear as 'natural' solutions to the problems we face" (ibid.). Yet Power also identified managerial fashion as a crucial vector of change. Thus, "just as other fashions have come and gone as the basis for management thinking, the audit explosion is also likely to be a passing phase. The seeds of a change may be there." (ibid.)

Yet perhaps the question of what these seeds are, and what they might grow into, hasn't been addressed sufficiently. Indeed, it might be viewed as somewhat surprising that Power's general diagnosis remains standing more than 15 years after he first suggested that the seeds for change might already be there. Here we suggest that one of the developments pointing "beyond audit" is audit implosion.

Wikipedia offers this definition of implosion:

> Implosion is a process in which objects are destroyed by collapsing on themselves. The opposite of explosion, implosion concentrates matter and energy. An example of implosion is a submarine being crushed from the outside by the hydrostatic pressure of the surrounding water.[6]

If *audit explosion* pointed to the extraordinary proliferation of audit initiatives across very numerous Western practices, *audit implosion* characterizes a situation in which auditing, as a particular form of control ("policing of policing") that is based on a specific ideal of "objective" knowledge, threatens to collapse under its own weight.

In the context of performance audits carried out by the National Audit Office, we have observed efforts similar to those described by Power, which aim to redesign the audit process "in the direction of a business service" (Power 2000, 2). In particular, audits are increasingly seen to fulfill a dual role of controlling audited organizations and creating opportunities for them to learn. Here we aim to capture some transformative dimensions of these efforts. We add specificity to the notion of audit implosion by characterizing some ways in which performance auditing operates through loops of control and learning.

Procedures for Making Knowledge

Our initial encounters with auditors generated surprise. We expected to meet a group of "dusty accountants" concerned mainly with keeping the audited organizations at arm's length in order to ensure the objectivity of their reports. These expectations were not met. Instead we found ourselves talking to people whose concern was how to carry out performance audits in a way that would inspire the audited organizations to internalize robust accounting practices.

It wasn't difficult to get in contact with auditors. One reason might be that, as a Danish public institution, the office has a legal obligation to respond to inquiries like ours. But auditors also exhibited a general willingness to discuss their tasks. Thus, we managed to set up initial interviews with relative ease. Nevertheless, the openness extended only so far. Our requests to observe auditors' daily work, join internal meetings, and participate in discussions with "clients" were politely turned down, and joining auditors on a "field trip" turned out to be impossible. Accordingly, our analysis is based on four open-ended interviews lasting between one and two hours, on documents we were given by the auditors, and on a study of the guidelines, brochures, reports, and other documentation that is publicly available on the Web. Using this material, we drafted a first version of the present chapter.

However, this turned out to be neither the end of our empirical studies nor the end of our empirical endeavor. For after we sent our informants copies of the draft, we were summoned to a meeting with the Head of Office and one of our "key informants." During this meeting we discussed not only empirical details but also analytic content and the "aesthetics" of our argumentation. The meeting, and several ensuing rounds of email discussions and revisions, produced additional data.

As auditors saw it, this process helped to correct our mistakes, prevent flawed generalizations, and clarify points of procedure. As we saw it, these interactions exemplified and mimicked the hybrid relationship between controller and counselor that we aimed to characterize in the first place. Specifically, the exchanges seemed to replicate the ways in which audit reports are themselves finalized in collaboration with audited organizations. This is a sort of participatory validation process in which drafts of reports are sent back and forth between auditors and audited organizations, which are encouraged to offer points of clarification and give critical feedback to be taken into account in subsequent revisions of the reports.

One outcome of this process is that most quotations from auditors' correspondence had to be removed out of concern that "colorful" expressions could be traced back to individuals. Because some of our interviewees were no longer employed at the office, we were told that it would be difficult to get their permission to be quoted. As we saw it, this was a somewhat protective stance that exemplifies the importance the office attaches to formal procedure rather than individual "opinion." As our contact persons saw it, the issue was one of research ethics.

The aforementioned dialogue has thus not only informed but literally track-changed this chapter. Recursively, the argument that we present has itself been subject to a version of the auditing practice it describes. This is analytically relevant because it indexes how maintaining a sense of control in situations in which the authority of the audit office is perceived as potentially at risk is still very important. In our particular case, the challenge of authority was handled "backward" with reference to all the formal procedures and guidelines that are in place to ensure that the office's products can be trusted (even in the context of an open and collaborative attitude).

This process had stylistic repercussions as well. Thus, one important effect of the removal of quotations and the toning down of any potentially controversial statement is the somewhat bland presentation of what follows. We urge readers to consider this blandness as indexing a kind of textual internalization of what the ethnography allows us to argue about performance auditing: the importance of retaining a neutrally, descriptive, or "epistemologically objective" tone of voice, even in, or especially in, a context in which auditing is increasingly influenced by new forms of partnership and attendant reflexive ways of making knowledge about organizations.

To some extent, then, the formal, procedural style we have just pointed to, and which this chapter replicates to an extent, stands in contrast to what we learned from our interviews about auditing as a highly reflexive and contingent endeavor. The more general point is that "procedural" or "formal" auditing and the participatory and reflexive endeavors we describe as *audit loops* co-exist and co-evolve. As our self-exemplifying case suggests, such co-evolution may create both epistemological and procedural friction. Our interactions with the office regarding both the form and the content of the present chapter highlights what we demonstrate in more detail below: that auditing works recursively through a series of loops of control and learning.

Approaching Implosion

In the introduction to *Audit Society*, Michael Power began envisaging a shift in auditing practice that he designated "audit implosion":

> [I]n the context of external financial auditing . . . it may be more appropriate to speak here of an "audit implosion" whereby organizations have become more 're-flexive' and where company directors have been forced to require responsibility for internal control systems and risk management. (1997, xviii)

The shift identified in this quote is one in which auditing ceases to be an externally imposed measure and increasingly becomes a matter of ongoing reflexive self-monitoring. In later work, Power continued to emphasize this transformation, arguing that "the growth of interest in internal control systems in the last ten years has accompanied an audit *implosion* meaning that auditing and inspection are now part of what organizations do to themselves" (2007, 162). Power related this change in particular to financial auditing practices. In a 2000 working paper titled Audit Implosion, he proposed to view the implosion of auditing as involving "an ongoing contraction in the significance of the external audit as a force in corporate governance, as compared to internal mechanisms and agencies of control" (2).

Power, however, is not the only social science scholar to make analytical use of the metaphor of implosion. The feminist theorist of technoscience Donna Haraway, too, characterizes implosion as a tool with which to understand socio-technical and organizational forms of "collapse":

> In the imploded time-space anomalies of late twentieth-century transnational capitalism and technoscience, subjects and objects, as well as the natural and the artificial, are transported through science-fictional wormholes to emerge as something quite other. (1997, 4)

Whereas Power's focus on implosion is rather concrete, having to do precisely with the internalization of auditing, Haraway makes it clear that implosion has far broader implications for cultural diagnosis. In particular, Haraway considers implosion to have emergent and unforeseeable effects on broader patterns of science, technology, and enterprise. This is why she uses the metaphor of "wormholes" from which objects and practices are likely to "emerge as something quite other" (4). Her analytical interest in unforeseen effects ties in with the anthropological attention to audit cultures, which, in Strathern's formulation, take particular interest in how "audit and ethics are structuring social expectations in such a way as to create new principles of social organization" (2000c, 281).

Of course, Power's identification of audit implosion understood as a "reflexive" internalization of audit practices exemplifies one "new principle of social organization." However, inspired by Haraway's argument that implosion heightens attentiveness to "heterogeneous and continual construction through historically located practice" (1997, 68), we suggest that the wormholes of auditing have a number of additional ramifications.

Audit implosion, in the Harawayian sense, has to do with things other than the internalization of auditing in organizations that have begun to institutionalize "second-order self-descriptions" (Strathern 2000d, 313). As we continue to discuss, this transformation centers on the emergence of increasingly complex auditing loops, a notion we use to index mutually shaping interactions between auditors and auditees that cross organizational barriers in multiple directions, both "downstream" and "upstream."

In this situation, implosion characterizes a situation in which both the epistemology and the form of auditing are put under pressure as external and internal modes of monitoring begin to work in tandem through auditing loops. In a broadly diagnostic vein, we end this chapter by considering the looping infrastructures of auditing as a response to a situation of knowledge production that Marilyn Strathern has referred to as "postplural."

Auditing Performance

The National Audit Office is divided into five departments.[7] One deals with administration, support, and international collaboration, two are in charge of financial accounting, and two conduct performance audits. The fact that the organization has a department dedicated to the auditing of performance as a type of auditing that is distinct from financial accounting is itself worth noting. The differences show in the organizational set-up but also in the different strategies defined for the departments. The strategy for the financial accounting departments aims to ensure the quality of the material based on which Parliament approves the state budget, to verify that the accounts of companies follow good accounting practice, and to set standards for good public accounting practice. These are rather unsurprising objectives.

However, the strategy for the department for performance audits looks quite different. Its goals are to prepare high-quality reports on the effectiveness of the administration to the Public Accounts Committee and to Parliament, to strengthen the development of effective administration by conducting cross-cutting and internationally based inquiries, and to inquire into areas in which administration runs the risk of not functioning effectively for citizens and businesses. The latter two points, in particular, are indicative of the explosive expansion of auditing into new areas.

Generally, the department for performance audits conducts two kinds of auditing. One is carried out at the request of the Public Accounts Committee and is mandatory. However, the department also initiates its own investigations. One of the recurrent tasks of the department is to carry out annual "risk analyses" of the ministries. On top of this, the department also conducts "strategic analyses." Inquiries begin with a pre-study—three months of intense preliminary investigations to determine whether it is worth the effort to plan and conduct a full audit.

A number of other reasons suggests why pre-studies make excellent organizational sense. The first relates to the fact that ministerial areas might be said to be rather broad and diffuse. If the aim of auditing were exclusively to inspect the annual budgets, as is the case for financial audits, perhaps this problem would be less pressing. But how to monitor and evaluate an entity like the Ministry of Foreign Affairs with a view to identifying areas with a risk of inefficient administration? Many simultaneous uncertainties are in play here, relating both to the question of where to look (the central ministerial office or the embassy in Vietnam; the internal monitoring unit or the communication department?) and what to look for (questions about meeting structure, collaborative process, strategic visions, incentives, or something quite different?).

The broad diversity in themes that can hold auditing interest is manifested in some of the final reports that came out in 2008. Titles such as "Development aid to Bhutan and Bangladesh," "Management of government IT projects," "The effects of the efforts made to integrate newly arrived immigrants," "The restoration of the railway infrastructure," and "The Danish Defence's EH-101 Helicopters," although by no means comprehensive, are indicative of the breadth and width of audit topics.[8] Since none of the employees at the department are, can be, or should be experts in such areas, it is clear that concerted efforts at the level of pre-studies help to contain the uncertainties about what is actually to be studied and why.

If the examples cited above suggest that the National Audit Office is operating in an environment characterized by audit explosion, its consequences are manifest in the perplexity auditors confront. If everything (or at least very many things) can be audited, and if the auditing outcome (administrative effectiveness or risk) is hard to measure and quantify, what then to look for and how to go about doing it? Paying closer attention to how auditors in the department deal with these challenges will allow us to begin specifying some implications of audit implosion.

Uncertainties of Auditing

As has been noted, performance audits can be instigated at the behest of either the Public Accounts Committee or the Auditor General (based on input and suggestions from the employees in the department for performance auditing). However, the department also spends considerable energy defining its own strategic focus areas, which guide later choices of audit themes and specific topics. At the time of pre-studies, many decisions have thus already been made. One doesn't conduct a pre-study of the Ministry of the Environment *at large* but rather in relation to a circumscribed set of strategic areas.

Among other things, strategic importance relates to what is referred to as "being at risk." There is a double specification at work: topics must be important for citizens or politicians *and* they must involve processes and decisions that are perceived as risky (that is, there must be concern that the organization may not follow legal rules and requirements). This broad requirement, however, isn't selective enough to determine focus areas. Each spring, a strategic process is initiated in the department in order to designate themes of special importance for the subsequent auditing period. Among other things, this process entails inviting various high-profile researchers to offer input to the strategy process. These researchers do not directly propose audits. Instead, they make suggestions and provide inspiration for the choices the office makes. Examples of strategic focus areas might include effective administration or (during the recent financial crisis) economizing.

The specification of these themes narrows the potential scope of auditing considerably. Still, because of their generality, the relation of these themes to the specific set of concrete questions addressed by any particular performance audit remains open-ended. This prodded us to look further into the grounds for choosing specific audit topics and questions rather than others. However, we were told that, aside from the circumscribing strategic decisions just mentioned, there are no general grounds for such choices. In explaining how specific inquiries are chosen, one interviewee invoked an astrological metaphor: auditors look for "signs in the sun and moon." We were quite surprised by this choice of imagery. Asking where one might look for such signs, we were told that interesting suggestions might be found in the media but also internally in the organization. For example, good ideas might be developed in conversation with colleagues from the regular accounting departments who also spend significant amounts of time interacting with audited organizations.

Thus, the "signs in the sun and moon" came from a variety of sources, which also convey a sense of the uncertain political climate in which the National Audit operates. Deciphering these signs entails paying attention to a heterogeneous set of potential factors, something that is necessary even when one is working within the strategic framework. Somewhat naively, we had expected auditing decisions to be guided by wholly formal, objective criteria. However, it became obvious that such criteria were not in themselves specific enough to determine particular, practical choices about what to audit and how.

According to the formal description of the process, performance auditing follows a rigidly standardized schedule. First comes the pre-study, which is initiated by written contact to the selected organization or agency. An initial meeting follows, which clarifies the reason for the audit, the topics of investigation, and expectations as to what materials must be made available, and which specifies contact persons. The next document that is produced outlines the overall and particular auditing goals, accounting criteria, and accounting actions; it also explains the procedure by which the audit will be conducted. The design is spelled out in detail in the "accounting tree," which specifies accounting criteria (at two levels), accounting actions, and what documentation is required. The "accounting tree" and the audit plan are then presented to management in a collaborative process that generates further input and comments on them.

The second phase is data collection and analysis. An accounting note is created; it summarizes the documentation and offers preliminary evaluations and comments. This note is submitted to the organization that is to be audited, which has a chance to comment on and qualify the evaluation. Another document is then produced and circulated. Eventually a final report is submitted to the Public Accounting Committee at a specified time and date. Subsequently it is made public. As should be clear, this is a very regimented set-up. At the time of our interviews, the pre-study was presumed to take one to three months and the data collection and analysis phase *plus* the series of notes and reports that follow between six and nine months.[9]

This formal mode of operation perfectly exemplifies what the historian of science and administration Theodore Porter refers to as an epistemology of "mechanical objectivity" (1995, 4)—that is, objectivity obtained by following the rules. Following the rules, Porter explains, provides a "check on subjectivity: they should make it impossible for personal biases or preferences to affect the outcome of an investigation" (4). However, Porter qualifies this ideal in two important ways. First, "following rules may or may not be a good strategy for seeking truth" (4). Second, "mechanical objectivity can never be purely mechanical" (5). Indeed, it is because mechanical rule following cannot remain mechanical, and because it may not always be a "good strategy," that auditors are required to be constantly vigilant, supplementing their formal rule books with "signs in the sun and moon" such as input from colleagues, the news media, and researchers. As we shall see, performance auditing in action is also dependent in important ways on qualitative, interpretive processes.

Going into the Field in Search of Relevant Context

According to auditors, two general types of information are available: written information and information gathered by auditors during audit visits. To explore how these forms of information work together to produce audit knowledge, we will discuss a recent audit of the Ministry of Foreign Affairs that focused on development aid activities in an African country.

To operationalize and narrow the scope of this audit, a number of sub-goals were defined in the "accounting tree." They included answering whether the development goals were being met, whether guidelines for monitoring and evaluation followed best practice, and whether knowledge about aid effects and results were being used systematically. Each of these questions had several sub-questions, and for each sub-question relevant documentation was identified. It included "Performance assessment framework," "Status reports," "Reviews," "Analyses from other donors," "Programme completion reports," "Review aide memoires," "Guidelines for monitoring," "Evaluation policies," "Administrative manual," "Correspondence and contracts concerning evaluation," and much else. As this lengthy list suggests, it is a basic condition of performance auditing that one works with very large amounts of documentation. In contrast to the experiences social scientists may occasionally have: that information relevant to their research can be hard to come by, auditors must deal with the question of how to delimit their material, since in principle they can get access to almost anything. The problem of what to look for is exacerbated by the fact that many of the audited areas, including the development aid sector, are huge "paper mills," and thus the body of potentially relevant documentation is always enormous. The tension embedded in the auditing process is that this material must be analyzed in such a way as to answer the pre-defined questions and must fit into an accounting note no longer than 30 pages. The question of how to condense material is thus omnipresent. Furthermore, interpretation is always necessary, since auditors have to estimate the quality of arguments used in these reports and the interests behind them. To explain why it is necessary to be on guard for ulterior motives cropping up in reports, an auditor conjured the example of a fictive report which argued that "everything went wrong but it wasn't the fault of the Ministry, it was all because of an external partner." Because it is always possible for reports to invoke external uncontrollable factors to account for problems, auditors can never take them at face value.

The illustrations show that audited organizations are themselves not neatly bounded; rather, they operate through interactions with many other partners. Because there are usually multiple stakeholders, and because any project's outcome is a result of interactions between them, there is rarely a single causal explanation for any particular problem. This is one reason why the auditing of performance raises issues of interpretation. To contain the problems of interpretive flexibility posed by such "multi-causal" complexity, auditing performance invariably requires making empirical and analytical "cuts" that allow the auditors to delimit the scope and relevance of information. This includes making choices about what is the right data, and it involves assessments of its reliability.

Nevertheless if something "looks peculiar" in the available material, and the data doesn't explain why that is so, auditors are faced with an interpretive gap. Sometimes this gap can be bridged by triangulation with other documents, but the oddness may remain. One reason why auditors go beyond written documentation is to strengthen their interpretive capacity. To do so they go on audit visits, during which, to be sure, they acquire even more documents but during which they also interview a variety of people, participate in meetings, and attempt to put together a coherent perspective on the mass of data. Audit visits thus help to assure the quality of documentary evidence as well as of their own interpretations and analyses. To some extent this is a matter of developing a "gut feeling" for what is going on in the field.

For example, during an audit visit to an Africa country, auditors perceived that they would have an opportunity to participate in the project's annual general budget support review meeting; accordingly, they scheduled their visit around that meeting. Such participation enables auditors to get a sense of the "backstage" collaboration taking place in a project—that which isn't visible at the surface level of documentation. For example, a report might "openly" acknowledge that certain issues are difficult to discuss at meetings, but rarely will a report spell out in much detail precisely how or why these issues cannot be discussed openly. Such statements therefore obscure as much as they make visible. By attending meetings, however, auditors can get a sense of context. What kinds of issues are discussed, and how difficult is it to discuss them? Observation of organizational activities thus provides auditors with interpretive anchors that they can use to evaluate the statements found in reports. For similar reasons, auditors also usually conduct interviews with a variety of stakeholders—for example, evaluators, consultants, embassy employees, and local collaborators. Although the extensive formal documentation analyzed by auditors provides the basis for the final report, it thus matters a great deal that an auditor be able to look people "deeply in the eyes."

"Field" experiences enable auditors to develop shared understandings of how to interpret the statements and actions of various stakeholders. At best they facilitate a "quantum leap" in understanding the situation. Further, even while auditors are focusing on the particular questions specified in the "accounting tree," they may perceive collateral issues which suggest that something strange may be going on in the organization. In turn, this may lead to supplementary queries, interviews, or conversations that allow the auditors to put their overarching auditing questions in a broader context. Thus, the "quantum leap" takes the form of a changing context for interpreting what are the issues and risks at hand.

These descriptions of auditors' "field work" demonstrate the extent to which qualitative, participatory methods have been invited into the auditing process. It was obvious to us that these methods to an extent resembled those used in our own research practice. But how did they form an evidence base for the performance audits? And how were qualitative data validated? Queried about the resemblance between qualitative audit knowledge and the interpretive analytics of social science, auditors readily acknowledged the similarity. They agreed that the process was based on judgment and estimation. Their pragmatic stance was that they had to "start out with what is there" and then decide whether the data was good enough. They also explained how, upon returning from "the field," auditors face the challenge of fitting their amorphous data, observations, intuitions, and interpretations into a report that must be "bulletproof" before it is made available for public scrutiny. One important way of ensuring validity is to be in close dialogue with the organization being audited, and to use that organization's critical comments to pinpoint weaknesses in the auditors' argumentation.

Yet in spite of the similarities, the methods, analytics, and outcomes of social scientific methods and auditing methods also differ in crucial ways. We have already noted that, at least in principle, auditors have "access to everything." Consequently, the auditors' problem is not usually how to get their hands on data, but rather to "define when enough is enough." In addition, whereas anthropologists and sociologists have recourse to theories and concepts that help them shape their questions and concerns, auditors relate their choices to "how the real world works." And because the facts of the case are presupposed, in the end, to speak for themselves, this analysis must be made entirely on the basis of guidelines and procedures, without relying on any guidance from concepts or theories. Only in this way can auditors ensure the "mechanical objectivity" of the audit process and, thus, the legitimacy of auditing.

Ensuring Legitimacy

The mandate of the National Audit Office expresses two aims: to ensure that good accounting practice is followed and to strengthen the development of effective administration. Accomplishing these two aims requires that auditing be trustworthy. Institutionally, the National Audit Office is kept separate from legislators, since it belongs to the judiciary. Despite this formal separation, there have been stories in the media in which auditors' judgments have been challenged and auditors have been presented as biased. In one sense, participatory methods such as fieldwork help to ensure the legitimacy of the final product because the auditors have been in the field and have seen what is going on. At the same time, though, basing judgments and recommendations on auditors' experiences in the field poses a threat to auditors' legitimacy.

During an interview, one auditor emphasized that audit visits are invested with considerable importance. Aside from the interpretive anchoring, on which we remarked above, he explained, audit visits can have a legitimizing function. He claimed that certain audit visits had made no difference whatsoever in the writing of the reports because all the relevant documentation was at home. The sole reason for traveling, in such cases, was to guard against the accusation by the organization under scrutiny that the report had been mere "desk work." In order to demonstrate accountability to the organization that is being investigated, apparently, auditors must show their faces. The broader point here is that, in order to be seen as trustworthy, auditors must, to an extent, be perceived as "on the same team" as those they audit, even if one purpose of getting closer to the realities of an organization is to inspect whether something fishy is going on within it.

But establishing trustworthiness by "stepping into practice" (Jensen 2007b) works only in relation to audited organizations. In order to ensure the legitimacy of audits to politicians and to the public, reference to having experienced real-world complexity is insufficient; it might even turn out to be detrimental. For to those audiences, what is important is precisely the reverse: to be able to demonstrate that the audit keeps "outside practice." Retaining distance, in these interpretive contexts, provides a guarantee of the objectivity of the evaluation.

For such reasons, auditors repeatedly emphasized that great care must be taken in preparing the final notes and reports so that there is absolutely "no room for error." Even though huge sets of data are gathered in preparation for writing the final report, most of the information never makes it onto paper. If one is only "90 percent certain" about information, it cannot be used. Spontaneously comparing auditing to research, one interviewee said that, in contrast with researchers, auditors have "no leeway for making mistakes." And because they have no leeway, they are required to always make conservative judgments.

In conjunction with the demand that each question and each subquestion defined by the "accounting tree" must be rigorously addressed, and with the fact that reports are relatively brief, the conservative requirement to avoid error at all cost determines the usage of information. The most immediate consequence is that formal quantitative documentation always forms the backbone of the evaluations offered in reports. Whereas it might be possible to construct a strong "accounting proof" based primarily on qualitative material, it would be "bloody difficult," since it would require a large-scale audit comprising surveys and interviews with, perhaps, hundreds of people. And even in such a case the issue of interviewees' self-censorship would have to be addressed methodically. Such auditing, we were told, would be considered only under very special circumstances.

This raises the question of how efforts to plan and conduct audit visits, participate in meetings, do interviews, and interpret qualitative data *matter* in relation to accountability. It also raises the question of whether knowledge made through such processes is an interesting but ultimately inconsequential "subjective supplement" to the quantitative information or whether it has a more substantial role to play. To understand how qualitative knowledge matters, one must pay attention to how the interpretive endeavor is rendered implicit and condensed in the final reports. In these reports, we were told, interpretations are conveyed "in the choice of words and the space given to particular topics."[10]

Thus, the data collected on the basis of qualitative methods has to be both absent and present. It has to be absent for the report to function as a legitimately objective evaluation in the eyes of Parliament and the general public. But it also has to be obliquely present for the qualitative knowledge gained from participatory methods to have a chance to guide the attention of those who know how to look, and who have an interest in inducing organizational change.

Learning from Auditing

The ambition to strengthen organizational control measures was central to the New Public Management initiatives of previous decades (Bloomfield 1995; Du Gay and Salaman 1992). Gradually, however, that ambition has been complemented with an interest in enabling organizations and agencies under scrutiny to *learn* from audits. The self-disciplining and internalization of the goals of auditing that Power describes as a result of the audit explosion can thus also be identified in the way auditors speak of what they hope will be long-term outcomes of their work.

When a final report is publicized, the ministry or agency under scrutiny is obliged to address each point of criticism. If problems with information technology have been indicated, for example, the ministry might respond that IT consultants have been contacted, or that new contracts are being negotiated. Few problems can be resolved within a few months, the auditors recognize, but organizations have to demonstrate that they take the criticism seriously by responding with a plan for how to address the issues. After a few years, a follow-up investigation will be conducted. Still, whether reasonable progress has been made often remains a matter of judgment.

But how can auditors know whether an organization has institutionalized ways of thinking and practices appropriate to the "accountable organization?" How can they detect whether learning has occurred? At stake is an experience that the National Audit Office shares with other ministries that are involved in ongoing evaluations. The Ministry of Foreign Affairs, for example, must account for the effects of the development aid projects and programs it funds—such as the ecotourism project we described in the previous chapter. As one auditor pointed out, although measuring output may be easy, measuring outcomes is often very difficult. For example, he said, the number of wells dug in the course of a development project (an output) can be counted, but how the wells have contributed to improved health in a region (an outcome) is much harder to show.

Nevertheless, performance audits must deal with outputs and outcomes simultaneously. In particular, in addition to evaluating outputs, they also have to induce organizations to internalize a specific "conduct of conduct" (e.g., Dean 1999; Larner and Walters 2004; Rose 1999) that is presumed to lead to better outcomes over time. In practice, this tension is managed by asking the audited ministry or agency to write a note that outlines its plans for responding to the identified issues. Even so, this formal feedback tells little about whether and how much the ministry has actually learned from the audit. It isn't easy to decipher, from such responses, whether an audit has led to a process of internal reflection that will result in organizational change or whether it is intended to "satisfice" (Simon 1957)—that is, to merely *signal* that demands and suggestions are taken seriously, while in reality the organization is doing little to address the underlying issues.

In addition, auditors are aware that their inquiries are imbued with a sense of threat that may run counter to their aim to help organizations learn. Ministries, for example, may be fearful of audits because they worry about negative publicity. Thus, auditors engage in consciousness raising in the form of attempts to convey to ministries that they ought not be afraid. One auditor suggested that organizations are generally too worried. Worry usually is unnecessary and even counter-productive, he felt, since auditors "are not gangsters out to kill people." Occasionally, however, there might be good reason to be frightened of an audit: members of the organization in question may not have been doing a good job. Still, this auditor also pointed to how the feeling of threat works against organizational learning. In his ideal auditing culture, organizations would *welcome* a "brutally honest report" and use it to move ahead. The difficulty of establishing such a culture, in his view, was bound up with the fact that audited organizations are enmeshed in a political system in which reputation and legitimacy tend to be seen as more important than organizational improvement.

The same auditor went on to argue that, from the perspective of the National Audit Office, goal achievement in a narrow sense isn't always the most interesting issue. If one wants to know how some organization is doing, he argued, there are many more relevant questions to ask than whether it has achieved a particular goal. At least as important, in his view, is assisting organizations preventively by helping them identify issues that may become urgent later. Such assistance ties auditors and organizations together in a continuous feedback loop, simultaneously shaping the knowledge and the concerns of both parties.

Assisting organizations is difficult also because organizational actors may have different ambitions or strategies that may themselves be in friction, or even mutually exclusive. For example, successful short-term financial management may impede efficient long-term sustainability of project goals. Consider the example of measuring the efficiency of the educational system. Schools usually don't wish to do everything as cheaply as is possible, because their main aim is to teach children to be skilled at reading, mathematics, and so on. Of course the ability of the educational system to provide those competencies to children can be measured. But that takes time and costs money, and meanwhile the schools must continue to operate. The sensible thing to do in this situation, according to this auditor, is to realize that objective measurement is a long-term goal, and the system will have to run for a while before there is anything useful to measure anyway.

Interestingly, this line of argumentation runs counter to the thesis that audit is *only* exploding. It suggests that it is senseless to channel a constant stream of resources into audits requiring efforts by people who ought to spend their time differently. Accordingly, auditors are keen to avoid having their work result in "some very fancy reports with no relation to any reality." Thus, learning from auditing is, in part, also learning about how to circumscribe and make reasonable use of auditing itself. Auditing, at least as seen by some auditors, should apparently not explode *too much*.

Audit Loops

The situations we have described in previous sections are indicative of a gradual change perceived by auditors themselves. This is a change toward what we might call proactive auditing; that is, auditing that not only will control organizations but also may lead to their enhancement. One consequence of the requirement that auditing should "add value" is that the National Audit Office becomes increasingly implicated with what it evaluates. Here we are entering the audit wormhole. Through processes of proactive auditing, the office and the audited organizations become involved in feedback loops of information sharing, recommendations, suggestions, and responses, not dissimilar from how employees of NatureAid Denmark interact in loops of knowledge making with their partners. Ambiguity in the relationship between the external and the internal may result.

"It is a balancing act," one auditor told us. Holding the balance is crucial, however, since politicians or journalists may jump at an opportunity to argue that the National Audit Office is less than objective.[11] "As soon as you touch these things," one interviewee noted, "you are in the political sphere." Since auditors, simply by doing their job, risk being drawn into the political sphere, it is crucial for them to avoid to be seen as taking sides. The auditors we spoke with insisted on objectivity, neutrality, and formal transparency even as they aimed to provide learning opportunities.

This is why it is so important to the National Audit Office that performance auditing is seen by the public and by members of Parliament as a straightforward evaluative process. Data flows from the organizations and agencies that are being investigated to the audit office. Suggestions and criticisms flow from the audit office to the organizations and agencies and back. Responses and clarifications are written. Recommendations are received and put to use. In contrast to this linear, procedural self-description, we suggest that the framing of auditing, the production of audit knowledge, the recommendations and demands such knowledge gives rise to, and the subsequent responses are recursively implicated. Through loops of ongoing interactions, these phases become "folded together," and the question of what constitutes the inside and the outside of auditing is rendered ambiguous.

The situation can be explicated with reference to the previously mentioned notion of mechanical, rule-following objectivity. As Porter notes, this form of objectivity is "especially prominent where inside and outside are not sharply differentiated" (1995, 229). This lack of sharp differentiation precisely characterizes the loop-like interactions through which performance audit knowledge is made and validated. This is the kind of situation, Porter argues, that "encourages the greatest extremes of standardization and objectivity, a preoccupation with explicit, public forms of knowledge" (229). Such a preoccupation is also a prevalent feature of performance auditing. It is manifest, not least, in the way the indisputability of final reports is seen as depending exclusively on "objective" data, even though the quality of such reports is acknowledged to rely to a significant extent on qualitative knowledge and observations made "in the field." Because of these co-evolving but also competing and (to an extent) contradictory tendencies, auditors encounter challenges when engaging in audit loops in support of organizational learning.

As has been noted, these are loops through which information continuously flows back and forth between auditors and auditees, transforming the knowledge of both parties in the process. Such audit loops, we argue, are *constitutive* of performance auditing, because they are absolutely necessary for the production of relevant audit knowledge. At the same time, these loops have contradictory implications. They are both generative and troublesome. They form the basis for both control and learning while also threatening both aspirations. And they involve interaction and mutuality in a way that cannot be accounted for according to the epistemology of mechanical objectivity that both informs public and internal understandings of auditing and sustains the organizational legitimacy of the National Audit Office.

Although we have illustrated audit loops with material from performance auditing at the National Audit Office, they are by no means confined to that setting. Indeed, audit loops swirl in and out of innumerable organizations. For example, internal auditing offices of large organizations collaborate with the National Audit Office as well as with other external consulting bodies concerning questions such as how to implement and improve their own auditing measures and practices of monitoring, their own infrastructures of accountability. (See chapter 5.) In addition, audit loops connect the National Audit Office with its clients, just as they connect the internal auditing bodies of these clients with those they monitor. But loops of information and knowledge making also extend in the opposite direction, as we discovered when we queried whether the National Audit Office is itself audited. It was in the context of addressing this question that we were told how auditors have "no leeway for making mistakes," since any mistake risks tarnishing the reputation of the organization as a whole. One could certainly argue that having (in principle) no leeway for making mistakes doesn't prevent an organization from making them. Yet it is to minimize any chance of making such mistakes that the National Audit Office enters into additional audit loops in order to ensure its accountability.

Thus, since 2000 the National Audit Office has asked external reviewers to evaluate the quality of all its reports. In 2006, the office collaborated with the Supreme Audit Institutions of several other countries in a process in which colleagues from these institutions conducted a peer review of the National Audit Office and offered criticisms and recommendations. The detailed report and the organizational plans for addressing the shortcomings that were identified were similar in format to those usually used by the National Audit Office for its performance audits. These documents are also publicly available online.[12]

Accordingly, we might say that even if no one forces the National Audit Office to "take its own medicine" it willingly does so anyway. It engages in exercises that help it perform as the kind of learning organization that it hopes to induce other organizations to become.

A Postplural Condition?

In preceding sections, we have used activities relating to the conduct of performance auditing to illustrative purpose. In particular, we have aimed to elucidate some conventions and cultural forms of auditing and explored some ways in which they may be changing. Inspired by Marilyn Strathern and Donna Haraway, we see such "seeds of change" as relating to a process of cultural transformation in which auditing is imploding on top of, or alongside, its explosions.

In *After Nature* (1992a), Strathern refers to a "merographic amalgam of pluralism."[13] Here pluralism is understood epistemologically, as indicating that many perspectives might always be taken on any given case. The notion of merography (a word whose root, *mero*, means *fraction*) points to the often- implicit pluralist idea that each perspective taken on each case adds to a general body of knowledge. In epistemological pluralism, knowledge is thus gained in a dual process. On the one hand, one gets to "know more" by examining more cases. On the other hand, one may "know more" by adopting new perspectives on existing cases.

In a formulation that resonates with the view from performance auditing that we have tried to establish, Strathern notes that "if the assumption is that much of what is invisible is what is simply *not yet made* visible, then there will always be more to learn about the organization, further realities to uncover" (2000d, 312). The proliferation of performance auditing and its ambition to delve into more and more aspects of the regularities of organizational life seem, indeed, to point in this direction.

Strathern's central observation, however, is that "a postplural world has 'lost' this mathematics." The term *postplural* designates a world in which epistemological pluralism is fragmenting, and, in a sense, collapsing under its own weight. On the one hand, this is a consequence of the empirical observation that one will never be able to exhaust the cases that can be investigated. On the other hand, it is a consequence of the analytical realization that the proliferation of perspectives embedded in methods, approaches, and questions do not add up, because they continue to open up to an indefinite horizon of inquiry that can have no endpoint. We might say that *the postplural* indicates a situation in which the suggestion that the more you know, the more you know that you don't know—and, indeed, cannot know—is taken to its logical end. This line of thinking provides us with some tentative indications of where auditing may be headed, and the looping characteristics of auditing suggest a direction.

As we have argued, auditors operate simultaneously on a terrain of control and a terrain of learning. This duality is operationalized in an infrastructure of auditing that takes the form of a series of interactive loops, weaving in and out of organizations, through which audit knowledge is made. Here we make the further suggestion that this looping infrastructure makes the question of perspective central to this navigational feat. This is contrary to how auditors see it; for as we have shown, their epistemological authority and their organizational legitimacy are premised precisely on their having no perspective, only procedure.

Nevertheless, perspectives are embedded in audit processes in multiple ways. There are views (for example, from research) that lead to strategies, there are signs that lead to specification of particular audit themes and areas, and there are choices that lead to even more specific audit questions. Finally, there are concrete decisions about what will count as the relevant datasets of formal documentation, and about how to interpret the relation of such documents to the mass of informal, qualitative knowledge, gathered from audit visits. Additionally, we have noted how the need to access deeper and in some ways more relevant knowledge about organizations has required the more or less *ad hoc* adoption of qualitative and participatory social science methods, shorn, however, of any explicit conceptual guidance. Rather than mechanically determined by procedure, auditing is thus also contingent—perspectival—in multiple ways.

In particular, perspectives proliferate because the making of audit knowledge engages auditors in complex loops of interaction with those they evaluate and those who, in turn, evaluate them. One consequence of this situation is that audit organizations become subject to tensions comparable to those present in the organizations they monitor. These loops, we have suggested, are nudging auditing into a kind of Harawayian "wormhole," from which it may emerge as a "quite other" form of knowledge-making practice.

Specifically, audit loops are indicative of a situation in which there are multiple auditing aims, which do not necessarily add up to a larger whole, and partners proliferate alongside the questions that may be asked of organizations. This is the sense in which, even as auditing may have exploded, as Power argued, it has also imploded. This argument doesn't involve any predictions about the disappearance of auditing as either a form of control or knowledge making. As Haraway reminds us, although "implosion of dimensions implies loss of clear and distinct identities," it doesn't entail any "loss of mass and energy" (1997, 69). And, obviously, auditing as a set of everyday activities show no sign of vanishing.

What does seem to be imploding is auditing as a cultural form of evaluation based on formal detachment and rigid procedural rule following and as an epistemology relying on "mechanical objectivity." The ideal of detachment is imploding in a situation in which auditing by necessity implicate organizations and auditors in a looping infrastructure.[14] And the ideal of mechanical objectivity is challenged in a situation in which a multiplicity of perspectival choices and decisions underpin audit knowledge; in which knowledge is constructed through relations rather than through an "objective gaze."

Even so, formal detachment and mechanical objectivity remain central to auditors' own conceptualization of their work and roles. Thus, an epistemology very akin to Porter's mechanical objectivity is upheld as the official ideal, though it doesn't characterize the practice of auditing very precisely.

In this sense formal and participatory auditing processes co-exist, co-evolve, and alternate at different moments in time. We see these simultaneously complementary and frictional aspects of auditing as generated by practical efforts to contain and manage the paradoxes of postplural knowledge making in a world still assumed to be plural.

Critical anthropological studies of auditing (e.g., Pels 2000; Shore and Wright 1997) have tended to view auditing as an especially predominant and efficacious structure of control. In Peter Pels' characterization, for example, "the stated goal of making the inner workings of organizations more visible goes together with a positioning of the audit process itself as an increasingly private and invisible expert activity" (2000, 142). Auditing, Pels continues, "actively constructs the environments in which it operates, making them more 'auditable' at the same time that it renders itself thereby invulnerable to its own failure" (ibid.). In this description, auditing becomes performative in a double sense. It constructs the environment in which it operates in order to make it more auditable. Since the organizational environment is, in each case, specified by monitoring procedures, the audit system closes on itself; thus, it becomes blind to its own failures and limitations. But then the failures of auditing simply call for more auditing. The very activity (and ideal) of auditing is rendered invisible in this process, and this is why auditing is made invulnerable to failure. In this context, Susan Leigh Star's question (addressed to designers of computer systems) "*quis formabit formatores*?" ("who will formalize the formalizers") (1995b, 101), becomes pertinent as "who will audit the auditors?" (Cf. Ciborra and Failla 2000, 124.)

But is auditing invulnerable to failure? Contrary to Brown and Lightfoot's (2002) finding that transparency requirements do not enter into the managers' "control room," where the really important decisions are made, the National Audit Office strives to make its own practices available to further scrutiny by making additional audit loops that will make the office *as auditable and transparent* as it requires other organizations to be. Contrary to the idea that the office provides the final link in a chain of accountability, the infrastructure extends both "forward" and "backward" from the office. Since the infrastructure consists of loops, rather than a chain, it isn't possible to reach a final instance of auditing at all.

If detachment is seen as a prerequisite for mechanical objectivity, this might be perceived as indicative of a situation in which auditing, rather than an encompassing form of objectivized control, becomes contextualized and relativized. But in a postplural condition, the stark oppositions between objectivity and relativity, between detachment and attachment, lose much of their purchase. Clearly, there can be no auditing closure in a situation in which audit knowledge depends on a looping infrastructure. Each new level of auditing simply creates more monitorable contexts and issues.

Presently, monitoring has become a powerful "descriptor seemingly applicable to all kinds of reckonings, evaluations, and measurements" (Strathern 2000b, 2). This chapter has brought audit loops into view as a conceptual tool for describing certain infrastructural transformations of state bureaucracy, occurring in the name of accountability and transparency, as results of the tentative but markedly increasing partnership orientation of monitoring practices.

In view of the power of monitoring, perhaps it is not too grand to talk about the activities, practices, and imaginations that it shapes as adding up to a *monitoring movement* simultaneously shaped by and enacted as infrastructure and partnership. In the final chapter, we grapple with the issue of how to simultaneously characterize the consistency and the variations of this movement in light of the divergent cases we have studied.

7 Monitoring Movements

In this book we have examined the problematic landscape of monitoring in which aid infrastructure, partnership, accountability, and transparency co-evolve. Polymorphous ethnographic engagement with monitoring practices has been central to this endeavor. Thus, we have aimed to characterize ways in which new ontologies of aid emerge through monitoring and evaluation, and through the making of public information, which used to be accessible only to development professionals and government officers. We have argued that today's development aid is, to a significant degree, constructed by information, standards, and collaboration formats alongside the "usual suspects": interest groups, development organizations, consultants, and other aid professionals.

Though we began our research by largely taking monitoring practices as "found objects," we quickly realized that monitoring refused to stand still, both empirically and conceptually. We refer to the landscape of monitoring initiatives and infrastructures characterizing development aid today with the overarching notion of the monitoring movement. To an important extent it is impossible to decide "how much" of the monitoring movement belongs to "the empirical" and "how much" to "the conceptual." The monitoring movement thus designates an "integrated" empirical-conceptual package with its own (limited) performative aspirations. These are aspirations to change the ways in which, and the purposes for which, infrastructure and partnership in development are imagined, conceptualized, and made.

On the one hand, the monitoring movement gives form to a multiplicity of practices. These practices differ in their hopes and aspirations, standards and formats, in their practices of partnership, and, of course, in their infrastructures. Monitoring is "an area of agency," "a site where global and local knowledge come together in a dynamic improvisation" (Hemment 2004, 236). Through such processes, aid infrastructures not only *connect* local to global sites of action; they *make* the local and the global, and these new localities and globalities, in turn, shape infrastructural arrangements. Monitoring moves around, and what monitoring means, what it entails, whom it involves, and how it is done all move in the same process. As this happens, initial (design) intentions and (political) ambitions for accountability and transparency are undone and redone.

On the other hand, monitoring also stabilizes relations and gives consistency to the practices we have described. Emphasizing the consistency of monitoring across sites doesn't mean that there aren't differences or even contradictions between these practices. Hence, we have followed a Deleuzian line of inquiry that aims to explicate precisely what makes heterogeneous elements cohere in assemblages without rendering them homogeneous (Deleuze 2007, 176–179).

Exploring consistency and coherence allows us to talk about heterogeneous monitoring assemblages as forming a movement. Thus, at the most general level, our analyses focus on collective efforts to achieve transparency and accountability in aid practices. In this sense, they can be seen as forming a movement similar to other social movements (animal rights, ecofeminism, and so on) that, while internally diverse, are engaged in "the same kind" of effort.

As a social movement, the monitoring movement has a certain kind of content: it takes shape around an agenda for increasing accountability and transparency and enhancing the effectiveness of aid by means of information infrastructure. Our interest in the diversity and consistency of these practices has led to analyses that characterize the monitoring movement as simultaneously one and many, neither wholly consistent nor quite falling apart, not quite a multiplicity, yet a set of practices and knowledges that hang together as a "vague whole" (Verran 2011).

Now the coherence of social movements is usually viewed in terms of substantive aims. According to a dictionary definition, a social movement is "a group of people with a common ideology who try together to achieve certain general goals."[1] Shared in the monitoring movement is the ambition to find infrastructural solutions for problems related to aid effectiveness experienced in the past that will work in the future.

However, we have engaged with these practices not only by paying close attention to their aims and ambitions but also by considering the material forms through which these aims are expressed and transformed. As Bill Maurer reminds us, forms are underwritten by "particularity and perpetual revisability: if we have a standard form for something, we can play around within the spaces on the form or at the margins. That play may lead us to want to revise the form itself" (2006, 91). Maurer's observation allows us to recognize "perpetual revisability" as a an important dimension of the monitoring movement: even though much of its content consists of standards and formats, its substance is subject to ongoing revision.

Maurer's characterization of form comes in handy because it is concerned with pattern but dispenses with the requirement to base analysis on pre-defined scales. In addition, the notion of a form can be understood at once aesthetically and organizationally (as in the technical and social forms emerging in the monitoring movement) and literally and materially (as in filling out forms). The latter aspect resonates with the paperwork and procedures integral to the making of infrastructures of accountability. (See chapters 5 and 6.)

Actor-network theory has described the relationship between stability and novelty in terms of "overflows" (Callon 1998; Winthereik and Langstrup 2010). Overflows occur because actor networks are unable to order all the different kinds of relations they make and because any network is also shaped by activities carried out by actors who are involved in other projects and networks. Similarly, in the context of development aid it has been noted that monitoring tends to create "new forms and processes of global civil society" (Lewis and Kanji 2009, 156; cf. Edwards 1994).

To get a perspective on what is "new" about the "forms and processes" of the monitoring movement, we turn to an interpretation offered by a development advisor and consultant.

A Story of Regress

"Development, as in Third World development, is a debauched word, a whore of a word. Its users can't look you in the eye." This statement is from "The Development Game," a paper by a veteran development consultant that was published under the pseudonym Richard Frank in *Granta* in 1986 and later reprinted in *The Post-Development Reader*. (See Frank 1997.) We were given a copy of the paper by an informant named Lars, a development consultant who was about to retire after a lifetime of service to a major Danish development organization. He thought it entertaining and liked its cynicism.

According to "The Development Game," development aid works because consultants and advisors are able to separate the official world from the real world. The narrator describes his perspective this way:

> You have to make a choice about the world you live in. Nowadays I live in the official world. The real world is infinitely complex and even the people, who are part of it don't understand it. And we are here for only four weeks, most of that in office

meetings. When you discover that the official world does not correspond to the real world, you can either accept the official version or make your own judgment. It's always best to take the government figures. That way you save yourself work and do not tread on anyone's toes that matter. We are here, after all, as guests. (Frank 1997, 271)

Only those development workers who recognize the separation between these worlds can stay sane. Failure to acknowledge it leads to a form of idealism that is represented in "The Development Game" by a young, inexperienced Dutch researcher who naively believes that the aim of development is to help the poor. In the long run, Frank suggests, such people don't have the "stomach for development" (272). In reality, Frank argues, development is primarily a way for the West to maintain a dominant position in the world. (Cf. Cowen and Shenton 1996; Rottenburg 2009.)

Our informant Lars enjoyed Frank's rough treatment of development. He told us that the "The Development Game" reminded him of experiences he had had in the course of his own career. In his opinion, Frank's description of development as a game in which it isn't possible to believe that what you do is good was largely valid. "Even the NGOs know it," he stated.

Yet our conversation with Lars began with a discussion of a very different text: *Partners in Development* (Pearson 1969), which, as we noted in chapter 1, was influential in putting partnership on the agenda in the late 1960s and the early 1970s. Lars had absolutely loved the Pearson report. He had come across it at a university. "We were a group of students studying with this professor who had worked in the UN previously," he recalled. "We were idealists." Lars remembered how impressed he had been with the participatory approach. He fondly recalled that its recommendations were down to earth even though the committee consisted of "top people."

What happened in between the Pearson report and "The Development Game"? How did Lars come to think of Richard Frank's cynical account of development as a more fitting description than the Pearson report's positive message about partnership and world community? One plausible explanation is that, looking back on a life devoted to development, Lars found it hard to see what had been achieved. In general terms, global inequality is greater than ever, and Lars seemed to find it hard to pinpoint exactly how his own labor had helped. But he went further. He said that young people working in the development agencies of today are no longer idealists, but are mainly concerned with financial matters. Increasingly, Lars argued, development employees are bureaucrats—specialists in indicators and financial models—who have little sense of life "on the ground."[2]

These specialists in numbers and administration are of a different breed than development workers of Lars' generation, who went into the field to solve practical and technical problems by working with people face to face.

Lars' account of how development has changed stands in glaring contrast with official accounts, such as the one embedded in the Paris Declaration, which believe that aid's effectiveness can be enhanced through partnership, accountability, and transparency, facilitated by new infrastructure. In particular, Lars alerted us to a way of thinking about current efforts to create transparency in light of a past in which effects could be witnessed more "directly" (through field visits) and knowledge was therefore, in a sense, less uncertain. Obviously, Lars preferred the world of *Partnerships in Development* to that of the Paris Declaration, even if he had ceased to believe in its viability.

In Lars' narrative, mutual accountability and partnership don't naturally belong together, "because accountability is largely about finances." This view differs from practically all the other accounts we have come across in this book—accounts in which accountability and partnership form a synergetic unity. Curiously, however, if we look more closely at it, we see that Lars' story of regress is based on oppositions that seem to undermine the partnership vision.

Lars' narrative contrasts the idealism of yore with the bureaucracy of the present. Previously, we were told, participation was designed with partners, but the Paris Declaration worked to bureaucratize partnership and turn it into a formula. Thus, Lars distinguished between earlier efforts to build capacity in fragile states and current attempts to support emerging markets. Surprisingly, however, Lars also defined the difference between then and now in terms of an increasing realization that there *really are* big differences between "them" and "us."

"We" (that is, development consultants from the Western countries) have technical skills. But more important than that, "we" are capable of advanced critical thinking and problem solving. "They" (that is, people in developing countries) are lacking in technical and analytical skills as a result of educational deficits that ought not to exist, because of decades of educational aid, but nevertheless persist. Owing to these deficits, actors in developing countries lack the capacity to handle the complex processes on which successful development depends.

In the name of partnership, Western development bureaucrats, who know nothing of development realities, transfer responsibilities to people in developing countries, who do not know what to do with them except to favor their own pockets, careers, and networks. In Lars' account, infrastructures of accountability are in one sense hopeful spaces, insofar as they create a meeting ground between "us" and "them." "We" can provide "them" with technical infrastructure and possibilities of sharing information in new ways. In turn, "they" will be empowered because freely available information spurs everyone's critical thinking.

Even if administrative responsibility has gradually been transferred to developing countries through a "good governance" agenda (see Abrahamsen 2000), for Lars the problem of knowledge in development has more to do with fragile infrastructures than with fragile states. Thus, Lars emphasizes the looming danger that authoritarian regimes will resist transparency. The way to solve to this "knowledge problem" is to establish partnerships in which mutual accountability is the desired norm. One derived consequence is the notion, discussed in previous chapters, that aid infrastructures should be for "everyone."

Infrastructures for Everyone

Of course, aid information infrastructures to ensure accountability didn't come into being with the Paris Declaration. As long as countries and organizations have endeavored to "develop" other places, they have needed to share information and thus move it through some kind of infrastructure. In addition, issues of what we currently call "accountability" and "transparency" have a long history in development aid. During the Cold War, colonies gradually gained political independence while becoming increasingly economically dependent on aid. Those were the times during which discussions about fragile states arose. The insides of developing states were impenetrable to development organizations. "Corruption" became a development keyword (Williams 1981), indexing the opacity of these newly formed governmental entities. Strengthening the capacities of the administrations of developing countries was a way of working toward "transparency" and "accountability."

Indeed, a Westernized form of bureaucracy came to be seen as an indicator of a modernizing state. "Modern" bureaucracy became a proxy for continued colonial presence. It suggested that a developing country was "on the right track," headed toward a modernity assumed to take only one form: the Western one (Rostow 1960). This is the story told in the Pearson report and reiterated by Lars: Western specialists were needed to build capacity to ensure "good governance." The Pearson report focused on partnership, but the level of engagement was almost exclusively centered on states and their institutions. Development infrastructure was seen primarily as an issue of information sharing "at the macro level," between donors and the bureaucracies of development states.

From the point of view of the infrastructures of accountability imagined by the Paris Declaration, and gradually emerging through the endeavors of initiatives such as IATI and AidData, the ambition is *still* to "better understand" what goes on behind the closed doors of ministries and local governments of developing countries. Overtly, this is solely a matter of knowing whether aid works effectively and making improvements. Covertly, as development consultants explain, such knowledge is also about donor competition and safeguarding influence in particular areas or regions. In one sense, therefore, new infrastructures of accountability are *still* meant for empowerment at the national level. It is in their capacity as tools for empowerment that technologies such as development assistance databases are offered to recipient states' bureaucracies. (See chapter 3.)

In another sense, however, the situation seems radically different. There is no longer any separate national level, since aid partnerships nowadays move laterally across all "levels," from the micro to the macro, from individuals to state organizations. As we have seen, emerging aid infrastructures aim to bring many more actors into the "development game." Works in progress, these infrastructures reconfigure partnerships and organizations. While drawing on old imaginaries, they also create new ones. Co-evolving with the "development game," they are also involved in changing the game. At once, they produce old wine in new bottles (the old dream of transparency achieved through new digital infrastructures and geocoding) and new wine in old bottles (the new dream of infrastructures for everyone enabled by the old fascinations with comprehensive aid knowledge).

A recent AidData newsletter brought to our attention the International Open Government Data Conference, organized by Data.gov, the World Bank Open Data Initiative, and the Open Development Technology Alliance and held in July 2012. This conference emphasized that "what really matters now is what people are doing with the data to improve and/or change things." At her blog "Wait . . . What?—Bridging Community Development and Technology," Linda Raftree reported an argument put forth by a "public policy entrepreneur" named David Eaves:

> [T]he conferences we organize have got to talk less and less about how to get data open and have to start talking more about how do we use data to drive public policy objectives. I'm hoping the next International Open Government Data Conference will have an increasing number of presentations by citizens, non-profits and other outsiders [who] are using open data to drive their agenda, and how public servants are using open data strategically to drive to a[n] outcome.[3]

Similar to other imaginaries situated at the inventive frontiers of aid infrastructure (those of Publish What You Fund or IATI, for example), Eaves articulates aid data as facilitating radically new forms of emancipation that do not operate at the level of national government. Indeed, this is a view that sees infrastructure not only as a vehicle for transporting knowledge about the "real effects" of aid but as *itself the most important form of aid*. Aid data is depicted both as what enables the solution of aid challenges and as itself the biggest challenge.

This collapse of means and ends is a prominent feature of the transparency visions we have encountered. In this context, infrastructure can no longer be seen simply as an inert platform for the exchange of development knowledge. Nor, of course, can the kinds of partnerships that aid infrastructures are meant to support be delimited to the national level. We might therefore agree with David Eaves that some important changes are afoot.

Yet the visions of infrastructures for everyone that we have come across in this book are also implausible. If, as we have shown, neither infrastructures nor partnerships form wholes, and if they produce new knowledge gaps in the process of filling old ones, then the consequences of the monitoring movement must be sought in another direction than one of smooth, evenly distributed, and therefore emancipatory information flows.

The contrasts between the agenda of the transparency movement and the concrete activities of organizations such as NatureAid and the National Audit Office vividly illustrate the diversity of practices and aspirations within the monitoring movement. While the latter organizations are engaged in challenging but very concrete monitoring efforts, aid infrastructure advocates and programmers at IATI, AidData, and Publish What You Fund engage in the equally challenging but altogether different task of facilitating "meta-monitoring." Their interest is in how to make aid information available to any actor or organization whatsoever.

This aim accomplishes the double feat of infinitely expanding the set of relevant information while simultaneously expanding the set of potential users of that information. This is why the agenda of aid transparency fits so well with the equally amorphous and extendable concept of partnership, which, lacking scale, seems to have the potential to bring everyone into the orbit of development projects through the use of information technology. Obviously, however, not everyone is actually invested in these information infrastructures. The question of who finds hope for the future in aid infrastructures cannot be separated from the question of what problems one hopes to solve through them.

Infrastructural Problems and Solutions

The contrast between Lars' story of regress and the stories of progress through infrastructure expressed in policy declarations, in transparency campaigns, and in efforts to promote new monitoring tools in development organizations highlights the extent to which the technological solutions of today are based on knowledge about yesterday's problems and challenges. They are also based on the hope that the future is open for improvement. Lars, however, doubted this hopeful interpretation. In particular, he was adamant that aid problems cannot be solved by technological design, since infrastructures contain only the kinds of knowledge that powerful development actors agree to share. Such actors may welcome the information that others make accessible, but will be much more averse to opening their own files. For example, Lars had been involved in a project whose purpose was to produce virtual maps of political decision processes in certain Southeast Asian countries. The project had achieved very limited success because the governments involved had staunchly opposed the form of transparency that the project required. Though Lars was proud to have participated in this effort of digital democratization, he doubted the general effectiveness of such initiatives.

At a certain level, Lars' critical perspective aligns with our acritical one. In view of the entwinement of organizations, technologies, concrete and idealized partnerships, and practices and visions of accountability in aid landscapes inhabited by very different actors, it is indeed difficult to maintain that new aid infrastructures are simply "solutions" to a set of already known problems. Yet the realization that infrastructures are unlikely "solutions" doesn't mean they are empty shells in which only the information of the powerful circulates. Rather, as we have discussed throughout this book, infrastructures embed scripts and images of both a past and a future, and of new kinds of users and publics with new kinds of interests and relationships. As we have seen, infrastructures also set in motion new activities, forms of organization, and practices. Indeed, aid infrastructures are themselves highly complicated actors that co-evolve with the "development game."

Our previous chapters have vindicated the suggestion of Paul Edwards et al. that infrastructure is at once an "all-encompassing solution and an omnipresent problem, indispensable yet unsatisfactory, always already there; yet always an unfinished work in progress" (2009, 365). On the one hand, infrastructure is an "encompassing solution," one already there, wherever we have looked—for example, in the cases of AidData (which built on the OECD-DAC's Creditor Reporting System) and NatureAid (which relied on global formats and standards from their headquarters). On the other hand, it is also, and equally, an "omnipresent problem," an "unfinished work in progress." This is why entities as diverse as AidData and NatureAid continuously put so much effort into getting their infrastructures to do the things that infrastructures ought to do but which they seem perennially unable to accomplish.

This diagnosis inclines us to be skeptical about the assertions by Publish What You Fund and the International Aid Transparency Initiative that new aid infrastructure will inexorably lead to multiple-win situations. The issue is not only that the wins are hard to measure and objectify. More generally, the problem is that there is no external position, no political or "epistemological Switzerland" (Duerr 1985, 93), from which such wins and losses can be evaluated neutrally.

This observation can also be used to shed light on some of the ambiguities that haunt the question, asked by Edwards et al., of "how to integrate or replace existing systems; how to handle local divergence from a desired norm; including how (or whether) to keep users from working around or outside the system" (2009, 366). This appealingly pragmatic question recognizes that infrastructures do not, indeed cannot, assume or require any single norm or mode of conduct. But while this point is certainly well taken, the fact is also that the actors engaged with the infrastructures of AidData, NatureAid, or the National Audit Office are *already* dealing with this problem, in practical as well as conceptual terms. Yet they are doing so in very diverse ways and for different reasons. This is why adding our own "solution" to this diversity is most unlikely to resolve these issues. However, there are other options for STS studies of infrastructure than producing such solutions.

Inhabiting the Monitoring Movement

In her pathbreaking analysis of English kinship, Marilyn Strathern offered a seemingly innocuous suggestion: "[S]uppose that this problem were also fact: suppose that this incompleteness were part of English kinship thinking." (1992b, 92) Paraphrasing Strathern, we might pose a similar question with regards to aid information infrastructures: What if we accept that the "problem" of infrastructural incompleteness is "a fact"? What if there are inherent, unresolvable tensions between infrastructural standardization and divergent forms of use? Rather than attempting to practically or conceptually "solve" this problem, our task would then be to understand the processes that lead, in multiple settings, to the replication of the "fact" of incompleteness. This isn't a wholly "impractical" analytical stance. For example, it might lead to the identification and characterization of a diversity of viable, productive, yet incongruent ways of inhabiting the monitoring movement. (See chapter 5.)

We have seen that non-governmental organizations, global campaigns, and state agencies gather forces to build infrastructures of accountability that can be used by ordinary citizens. Partnerships between NGOs, donors, and other state organizations coalesce around the ambition to make aid transparent. Greater transparency will certify that aid money is well spent. It will also make clear who benefits from aid. Indeed, various digital platforms even become aid in their own right, since these technologies are meant to empower recipients, building their capacity to engage with donors on equal terms. (See chapters 3 and 4.)

We have also seen the establishment of quite surprising partnerships. Recall, for example, the group of political scientists who, aiming to do better statistical analysis of aid effects, ended up developing AidData in collaboration with a wide variety of partners. The project began as a research project, but it ended up as an alliance of scientists, designers and system developers, a wealthy philanthropist, private foundations, non-governmental organizations, and the Development Gateway, not to mention a set of imagined users.

Gathering at a pub after a long day at the AidData conference, project members narrated the origin story of the database. A PhD student's research project had run intro trouble because of the low quality and limited accessibility of aid data. The student and his supervisor concluded that it might be possible to dig up the missing data from various sources. Gradually more people and organizations, from the Development Gateway to the Hewlett Foundation, became interested. It became possible to envision a scaled-up project that would not only locate existing dispersed datasets but merge them through a digital platform, which might eventually also help to generate new information. At the point where we enter the story, the project was already very successful; while discussing aid data and ethnography at the pub, one of the principal investigators of AidData received a phone call announcing that the project had received another million dollars from a private funder. This story offers a good example of the tangled processes whereby actors come to "take up residence" (Kockelman 2010) in the monitoring movement.

Monitoring moves both spatially and organizationally. The AidData project grew by being drawn into an expansive and expanding assemblage, tying together the concerns of development practitioners, business stakeholders, governments, policy makers, and transparency campaigners. The PhD student's limited problem with missing aid data both connected with and was transformed by other actors' concerns with massive aid development challenges, such as corruption, transparency, and accountability. Thus, the aims, definitions, and aspirations of the people related through AidData run in many directions. People affiliated with AidData end up in the offices of the Development Gateway, where they work to create infrastructures for everyone. (See chapter 4.) The monitoring movement modulates through such ramifying relations.

The practices of NatureAid allow us to glimpse a quite different way of inhabiting this movement. On the one hand, the formal monitoring requirements of this NGO, like those of many others, have a certain built-in linearity. They entail a belief in monitoring made objective through formal procedure. This is a type of monitoring that purports to measure the world without affecting it. On the other hand, we encountered more reflexive and performative perspectives on the qualities and merits of monitoring. For the conservation team, monitoring and evaluation practices should facilitate mutual learning, but they should also perform the interdependence of partners.

During the midterm review discussed in chapter 5, one of the Danish consultants held a meeting with the head of a newly appointed M&E unit of a regional development program. The head of the unit showed considerable professional dedication and had great visions for how monitoring and evaluation would be able to provide the NatureAid network with better mechanisms for governing its activities. During the meeting, the head of the M&E unit argued that collecting more detailed information about individual projects would enhance the ability of the regional program to reach its goals. The Danish consultant was at once drawn to and taken aback by this ambition. He regained his composure only as he was presented with an enormous spreadsheet containing more than 1,700 indicators projected onto the wall of the meeting room.

Confronted with this spreadsheet, the consultant began to question the feasibility of working with such a massive system of indicators. He inquired about how this vast quantitative system would help his concrete efforts to do adaptive project management and support project partnership. The explosion of indicators promised to precipitate a large amount of extra reporting without offering any concrete benefits in return. Abstractly similar, the monitoring ambitions and activities of these two employees of the same NGO diverged as soon as the indicators became a practical issue. Clearly, monitoring exists in multiple versions even within this single organization.

Consider, finally, the case of performance audits conducted by Denmark's National Audit Office. (See chapter 6.) On the one hand, auditors take for granted the objectivity of formal procedure. On the other hand, they also make use of qualitative, participatory approaches, which they work to integrate with existing formats. Hesitatingly, they are taking on a role of partner on top of, or alongside, their traditional role as controllers. This office spends tremendous energy examining audited organizations in order to render their practices transparent. While doing so, auditors reflect on how to achieve a "good balance" of information sharing, evaluation, and control. Even if control remains the auditors' main task, monitoring also modulates in performance audits.

These movements and modulations are both organizational, engaging auditors in loops of control and learning, and conceptual, requiring auditors to attend to what it means to do auditing, balancing on a knife's edge as they weave together participatory engagement and procedural objectivity. Audit loops point to a blurring of the organizational and epistemological boundaries between monitor and monitored. We have argued that they are also indicative of an ontological transformation in which auditing no longer only explodes but also, simultaneously, implodes. The National Audit Office illustrates another mode—tentative, and in its own way even experimental—of inhabiting the monitoring movement.

As these various cases show, the monitoring movement is characterized by tremendous variations in practices, but also by significant overlaps in what matters to them. Still, these are irreducibly heterogeneous configurations. Actors and organizations inhabit the movement in vastly different ways. They even aim to define the possibilities of habitation very differently. Though the "end products" might be abstractly envisioned as aid infrastructures for everyone, there is little reason to think they are headed toward one final outcome, one coherent infrastructural whole.

Dividualization and Partial Knowledges

Forming no coherent whole, monitoring movements do not respect the boundaries and categories on which both proponents and opponents of the transparency agenda rely in their visions and arguments. In a context in which partnership loops and recursivity prevail, development cannot be understood in terms of linear, one-way processes of change—for example, as a general movement from state-centrism to market-centrism. Recognizing the recursivity of emerging aid infrastructures entails considering what monitoring movements may mean for the knowledge and the politics of aid.

Advocates of transparency in aid aspire to strengthen the links between technology and politics. This ambition is not itself new (see Beer 1981), although, as we have seen, some of the particular political and infrastructural practices are. The aspiration to ensure transparency through technological means can be related to Gilles Deleuze's diagnosis of societies of control, which focused on the repressive effects of free information flows (1995, 177–83).

Whereas advocates of transparency in aid want to make information dynamic in order to *liberate* people, Deleuze saw real-time flow of information as the ultimate source of societal *control*. If the ideal of the aid transparency movement can be said to be cybernetic in that it aspires to a situation of real-time policy response to real-time aid data, this aspiration bears an uncanny resemblance to what Deleuze characterized as "ultrarapid forms of flee-floating control." For Deleuze, as for Aneesh (2006), code is central to this change:

> [W]hat is important is no longer either a signature or a number, but a code: the code is a password. . . . We no longer find ourselves dealing with the mass/individual pair. Individuals have become 'dividuals,' and masses, samples, data, markets, or 'banks.' (Deleuze 1995, 180)

As this quotation suggests, Deleuzian societies of control obviate the dualism between individual and society on which traditional conceptions of (liberal) democratic politics rely. Using the same term with which Strathern (1988) described Melanesian forms of "subjectivity," Deleuze referred to this process as *dividualization.*

The term *dividualization* refers to a situation in which the Western notion of the autonomous individual holds little relevance. In the Melanesian case, this is because people are "integrally implied" (Wagner 1991, 163) with each other through social relationships. For Deleuze, the individual, so dear to liberal philosophy and politics, is washing away in a sea of data. Agency is dividualized because identity and knowledge is increasingly made up of—literally assembled and reassembled from—fragmented elements collected from "samples, data, markets, or 'banks.'" What can knowledge and politics mean in a situation in which recursive partnerships and looping infrastructures dividualize people and organizations?

Deleuze introduced the term *modulation* in order to talk about the agency of dividuals. In musicology, modulation marks a transition from one key to another, a transition that may be temporary or permanent (but no more permanent than the piece of music itself). As used by Deleuze, *modulation* conjures an image of a society that must be constantly reconstituted but remains in flux.

The notion that political efficacy might be gained through infrastructures that facilitate globally "coded and distributed presence" connects with the idea of modulating agency. It is part of an ontology in which development agents are distributed and increasingly respondent to changing information patterns rather than to knowledge gained in local organizational settings or "face-to-face" encounters. Operating through aid information infrastructures, we suggest, "the individual" and "the organization" turn into modulating entities.

However, what follows from this diagnosis is quite uncertain. As the advocates of transparency in aid see it, political and informational liberation is at stake: once the infrastructure is fully workable, people will be able to gain access to and act on all relevant information about any aid issue. Practical ontology, however, leads to a quite different perspective. For what we have found is not a set of practices converging on shared and transparent information. Rather, we have encountered recursive partnerships and weedy, looping infrastructures that pose continuous challenges and dilemmas for *all* development partners. Not only do these partnerships create more knowledge and capacity; they also, simultaneously, create more knowledge gaps, which circumscribe and redefine discretionary capacity.

This "post-plural" (Strathern 1992a), "fractal" figure of "simultaneously more and less," "better and worse" is vivid in the case of NatureAid consultants' seeking to know the realities of projects and the "real" opinions of their Vietnamese partners. It is also clearly evident in the cases of the National Audit Office (which attempts to get to the bottom of ministerial realities through close collaboration with auditees) and AidData (which has collected an enormous amount of data but struggles to figure out to whom the availability of this information ultimately matters). On the basis of these diverse cases, we suggest that, although the monitoring movement generally entails efforts to make more information available to more diverse actors, this availability is quite unlikely to lead to more coherent and holistic forms of aid *knowledge*. Information from multiple sources proliferates, but organizations' and individuals' capacities for validation, comparison, and interpretation are not similarly enhanced and intensified. With development professionals and all those involved "regular citizens" swimming in a sea of information and navigating a huge set of potentially relevant partners, overview is likely to remain impossible.

This problem didn't emerge with cutting-edge infrastructures. Thomas Richards argued in *The Imperial Archive* that the challenge faced by English bureaucrats bent on knowing colonial realities was "that facts almost never added up to anything. They were snippets of knowledge, tiny particularized units responsible for our current idea of information. It took a leap of faith to believe that facts would someday add up to any palpable sum of knowledge, and that faith often took the form of an allied belief in comprehensive knowledge. Comprehensive knowledge was the sense that knowledge was singular and not plural, complete and not partial, global and not local, that all knowledges would ultimately turn out to be concordant in one great system of knowledge." (1993, 6–7) To an important extent, this characterization still holds. Although the pool of development information, made available to increasing numbers of people, is growing rapidly, it doesn't add up to a comprehensive set, "one great system of knowledge." Accordingly, it still requires a "leap of faith" on the behalf of transparency advocates and app designers to insist that their efforts will solve development' knowledge-sharing problems. Because of the presence of advanced information infrastructures and the availability of increasing amounts of "snippets" of information, however, faith in the powers of information has grown rather than diminished. The "leap" required to believe in "one great system" of transparency is perhaps smaller than ever before. Yet, we have suggested, the layers and loops that enable the production and presentation of this information also create the very gaps and frictions that prevent the fulfillment of these infrastructural promises. This "problem," as we have argued, is "a fact" about emerging aid infrastructures, and perhaps any infrastructure.

We are, therefore, not arguing that the monitoring movement threatens to undermine more encompassing and holistic forms of knowledge that were previously available. As the quotation from Richards suggests, in development there have never been such holistic forms of knowledge. Instead, we surmise, transparent information turns aid recipients, project workers, programmers, and development researchers into dividuals, whose capacities for informed action are simultaneously enhanced and reduced.

Even as more information is potentially available from which to compose a "truthful" picture of aid, the need to select, sort, and organize this information introduces new forms of partiality and opacity into aid knowledge. The point resembles Alvin Toffler's (1970) notion of information overload—the idea that too much available information renders decision making impossible. To this observation we add another: Insofar as people and organizations are increasingly dividualized (that is, produced in distributed and fragmented versions by information infrastructures), the very notion of a singular and coherent agent presumed by information overload is increasingly challenged. In this case, dividualized users of aid information will not have been deprived of political agency, but the nature of this agency will have changed.

The aid infrastructure toward which AidData and Publish What You Fund work thus instantiates an ontology that translates politics through technology and produces modulated agency. At the same time, however, knowledge drawn from such infrastructures is gaining in legitimacy precisely because they purport to deliver transparent information. On one side, if aid infrastructures are *unable* to create the desired transparency because the multiplication of information creates its own forms of opacity, there is little reason to believe that new aid practices will be radically more accountable and efficient or emancipatory because an infinite amount of information is rendered accessible to an infinite set of potential users. Yet this also means there is no reason to see these efforts as leading to—or exemplifying—a society of control. Dividualization in the monitoring movement is not inherently bound up with Deleuze's grim vision, because the question of who or what controls what is itself mutating when even the knowledge practices of power organizations show signs of implosion.

Thus, our depiction moves decidedly away from any black-and-white perspective on accountability and transparency. Instead, an interest in the practical ontologies of the monitoring movement leads us to focus on recursive partnerships and emerging infrastructures, on complex practices giving rise to modulated agency, on the dividualization of knowledge and power, and on varied, disjoint efforts to inhabit worlds of aid development.

The Habits to Be Acquired

In chapter 1, we identified aid information infrastructures as part of a problematic landscape: the landscape of the monitoring movement. In the subsequent chapters, we explored this landscape by means of an approach that sees infrastructure as giving rise to new ontologies of aid.

Practical ontology entails at once a description of the work done to make the new, informationalized worlds of development aid by heterogeneous actors (the many people, technologies, and organizations engaged in monitoring movement) and a description of the work done by researchers trying to elicit that work. Practical ontology doesn't open the gates to an imaginative playground in which the world can be shaped into anything the researcher might think of. Rather, it is indicative of an extension of the empirical to encompass everything (Gad and Jensen 2010), including concepts and discussions that the researcher cannot avoid bringing along when engaging with the world.

Certainly, as ethnographers of science and technology, we can know about knowledge and practice only if we explore the ontologies that actors (human and nonhuman) are operating within, putting together, exploring, and experimenting with. People's imaginations, their discourses and concepts, and indeed their "epistemologies" are as much part of changing these ontologies as their technologies, their tools, and their organizational formats. If imagining, designing, and working with aid infrastructures is also, literally, engaging in world making, then the intellectual habits to be acquired must be attuned to the composition of multiple, partially related forms knowledge, practice, and politics in these emergent worlds—and to their divergent, surprising consequences.

In *Thinking with Whitehead: A Free and Wild Creation of Concepts*, Isabelle Stengers talked about "the habits to be acquired" according to Whiteheadian cosmological demands (2011b, 274). These habits are equally congenial to the study of practical ontology. Strikingly, Stengers' specification of these demands returns us to the imagery of weeds. Contrary to our usage, for Stengers weeds is not a metaphor for the built-in tensions of infrastructure. Rather, she writes of weeds as a metaphor for the generative possibilities of thinking:

> Cement rejects the interstices in which the weed grows that will one day crack it open. The dry stones, for their part, can of course be displaced, and the weed that displaces them might certainly be tempted to abstract from the fact that without the stone the earth in which it grows would not have held. But the wall of dry stones is not defined against the interstices; the latter belong to it just as much as the stones that make it up. (274)

Stengers distinguishes between walls made of cement and walls made of stone. Cement walls rejects the spaces from which will grow the weeds that are bound to destroy them. In contrast, walls of dry stones are "not defined against the interstices"; rather, the stone wall and the weeds belong together—they co-evolve.

The point of Stengers' metaphor is that creative analysis requires attention to and care for the interstitial spaces of thinking. In disciplinary terms, we might refer to the weedy spaces between STS and social and cultural anthropology, or between development studies and infrastructure studies. More generally, we might think of the walls that compartmentalize and separate the empirical and the conceptual, the analytical and the methodological, or, indeed, knowledge and practice, or politics, with little attention to what grows in their interstices.

As in infrastructure, weeds are always already there, and so, Stengers suggests, they "do not need to be deliberately cultivated" (2011b, 328). But one has to be willing to look for them. This is the case, Stengers argues, "even if what lurks in the interstices has no significant consequences as long as the social order maintains itself" (331–332).

We have argued that in current practices of development aid social order and technological order co-evolve through the monitoring movement. Surely such co-evolution occurs in other settings too. In a context in which STS, infrastructure, and development researchers are increasingly urged not only to identify problems but also offer solutions, practical ontology has ushered us onto a different path.

In the monitoring movement, order is maintained through modulation and transformation. This is the point at which we situate the performative aspiration of this book. Practical ontology enables us to focus on the weeds that both undermine and sustain infrastructures of aid. We suggest that it is important to devote more attention to the interstitial spaces in which weeds grow. Recognizing the importance of such spaces for the making of infrastructures contributes to understanding and appreciating the multiple, divergent ontologies that co-exist in the monitoring movement.

Notes

Chapter 1

1. The exception is David Mosse and David Lewis' edited volume *The Aid Effect: Giving and Governing in International Development* (2005), which contains extended discussions of partnership and accountability. Even that book, however, doesn't touch upon infrastructure and transparency.

2. Source: http://st-and-dev.net/sts/.

3. Edmund Leach would characterize development anthropology as "a kind of neo-colonialism" (cited in Ferguson 1997, 166).

4. For further discussion of the logical framework and its use by NatureAid, see chapter 5.

Chapter 2

1. These interviews were carried out in Hanoi in October 2008.

2. From "Cambodia, Laos, Vietnam. South-East Asia joint initiative. Mutual accountability" (consultant's report, 2008).

3. A British consultant working for Danida in one of the largest and most influential Vietnamese ministries offered a description of how partnership works. Partnership, he suggested, entails the continuous negotiation of one fundamental tension. On the one hand, there is the donors' concern: How much can you spend and on what conditions? On the other hand, there are "weak institutions," that are not really geared to work with the accountability measures defined by donors. For this reason, funds are often not transferred to ministries in a continuous, well-planned process, but are rather tipped into treasuries "in huge piles." This means that no one is ready to take action even when funds are actually available. Our informant argued that the reason this problem still hasn't been resolved is that there is a "hidden agenda not to do anything." In some sense, the existing system works. But it works by maintaining a kind of deliberate opacity that preserves the autonomy of the participants.

4. Source: http://blog.aiddata.org.

5. To make matters more complicated, Brit is a dance partner with one of the villagers, while our contact person in NatureAid turned out to be our research partner insofar as he produced documentation used in this book.

6. This material includes interviews and observations of Danish, English, Australian, and Vietnamese development consultants; interviews with representatives of the OECD, Danida, the UN, and the World Bank; and interviews with Danish and international development researchers, representatives of NGOs, representatives of various aid transparency initiatives, and Danish national auditors. Last but not least, it involves a year of fieldwork in the Danish branch of a global environmental NGO that had invited us in to study its monitoring and evaluation practices. We refer to this organization by the pseudonym NatureAid.

7. For some examples, see the report on the Seeking NGO-Donor Partnership for Greater Effectiveness and Accountability workshop held by the Inter-American Development Bank (source: http://idbdocs.iadb.org), the "Agency accountability framework" developed by the Canadian International Development Agency (source: http://acdi-cida.gc.ca), and Matthew Winters' paper "Accountability, participation and foreign aid effectiveness" (2009).

8. Reading *The Network Inside Out* (Riles 2000), it is hard not to feel that the same must have been the case for Riles.

Chapter 3

1. This happens by joint assessment of "progress in implementing agreed commitments on aid effectiveness, including the Partnership Commitment (Indicator 12)" (source: http://www.oecd.org/development/aideffectiveness/34428351.pdf).

2. Source: interview with independent development consultant. Thus, it is a source of frustration to development practitioners that the principle of mutual accountability appears most important, yet least operational.

3. Source: http://www.publishwhatyoufund.org.

4. Source: http://www.publishwhatyoufund.org.

5. Source: http://www.access-info.org.

6. Source: http://www.publishwhatyoufund.org.

7. The particular heading has been changed, but as of this writing the argument is still available at http://www.publishwhatyoufund.org.

8. Source: http://www.youtube.com.

9. The conference's webpage, available at http://aiddata.org, offers further explanation.

10. Source: http://www.telegraph.co.uk.

11. The word *informatted* is derived from Shoshana Zuboff's 1989 book *In the Age of the Smart Machine*. We use it to point to how present-day bureaucrats are changed by engagement with information technologies into agents who, ideally, should produce and handle constant streams of information.

12. In addition, the wired-up user on the ground resonates with other recently invented publics, such as the inscribed users of microloans and other actors at the "bottom of the pyramid." Such a user isn't necessarily in need of empowerment; he or she may already have many unrecognized resources that can be tapped into.

13. After the change in ownership, the site was available at http://dad.mpi.gov. It is no longer available, and we have not been able to find any online reference to a functioning webpage. The disappearance of this site, however, corroborates our point.

Chapter 4

1. Source: http://en.wikipedia.org.

2. This approach contrasts and complements the work of Madeleine Akrich (1992), who argues in favor of opening the black box of stabilized technologies in order to understand the political scripts on which they rely.

3. Source: http://www.youtube.com.

4. Here we concentrate on the demo for the PC version.

5. Source: http://www.youtube.com.

6. For critical discussion of the stratification of the Internet as a common space through user-generated content, see Christian Fuchs' work at http://fuchs.uti.at.

7. Source: http://www.youtube.com.

8. Source: http://appsfordevelopment.challengepost.com.

9. Source: http://www.ictworks.org/network/ictworks-network/836. The winner of the first prize was StatPlanet World Bank, an app designed by the company Statsille, based in Melbourne, Australia. Like the Development Loop, StatPlanet World Bank allows users to "visualize and compare country and regional performance over time." Other prizes went to the French Development Timelines and the German interactive quiz Yourtopia.

10. Source: http://data.worldbank.org.

11. Source: http://techpresident.com.

12. Source: http://www.ted.com.

13. Source: http://techpresident.com.

14. Source: http://appsfordevelopment.challengepost.com.

15. Source: http://www.youtube.com.

16. Source: http://data.irtheoryandpractice.org.

Chapter 5

1. Another group in the Conservation Department focuses on the management of natural resources in Denmark and other EU members.

2. The tension between the monitoring ideals of the Danish team and a "desk-like approach" reappeared during a meeting at which the M&E officer enthusiastically explained to one of the Danish NatureAid consultants how his unit wanted to collect enormous amounts of information. Struck by the under-exploited opportunity of collecting information about ongoing projects based on indicators, he wanted to build a "dashboard" for better monitoring. The Danish NatureAid consultant insisted that this approach, however wonderful in theory, would never help him or any of his colleagues in addressing challenges on the project level.

3. The Logical Framework Approach is an approach, adopted by Danida in 1989, that is used in the planning and management of all development projects and in all projects receiving support. In 1996, the LogFrame was revised to accommodate the new participatory ambitions. As part of the revision, it was argued that the approach had become a straitjacket for project management, and a more flexible use was recommended. The current approach "intends to provide practical, flexible tools for analysis and dialogues throughout the project cycle in order to foster commitment and effective project execution" (source: "Logical Framework Approach: A Flexible Tool for Participatory Development," at http://amg.um.dk). It requires specification of development objectives, immediate objectives, output, activities, and input for projects in a matrix, to allow easy overview of a project's success.

4. "Key species" are large mammals—tiger, rhino, ice bear—whose existence in a specific area indicates a rich fauna in that area. Often they are indicators of biodiversity.

5. In spite of the name, such assistance doesn't have to be technical in any literal sense. It can just as well focus on social technologies of participation, or capacity building, or it can entail offering advice on how to set up and manage a project.

6. This system takes the format of a planning cycle and includes a number of reflexive questions about how well the project progresses and about the relevance of outcomes and outputs.

7. As Cooper and Packard note, this is too simple: "We tend to think of NGOs as a category whose mere existence shows that 'civil society' is working to counter the state's dominance of development initiatives, but we also tend to treat those organi-

zations in generic terms, not exploring their varied ideologies, organizational forms, and relations to state mechanisms." (1997a, 29) Below we examine in detail the "varied ideologies" and "organizational forms" of one environmental partnership.

8. Cooper and Packard relevantly note that "the historian's or anthropologist's concern with context and complexity is neither more nor less separable from a self-serving professionalism than the development practitioner's concern with the replicability of project design" (1997a, 26). It is noteworthy that our development professionals are simultaneously obsessed with replicability *and* with context and complexity.

9. Or in the bureaucracy of NatureAid Vietnam, where the project is recontextualized once more: the experience with building the long house is generalized as "a community model for development."

10. We say this well aware of the criticisms raised against "naturalizing" social relations by analyzing them in organic or biological terms. The specter of social Darwinism and evolutionary psychology looms large. Of course problems arise when organic imagery is literalized as description of sociality. Yet perhaps social critics have been too quick to naturalize organic imagery as inherently reductive. Here we simply note that many areas of natural science and biology currently have understandings of complexity and emergence that match or exceed the imaginaries of social scientists when speaking of complex social relations. STS and social anthropology might fruitfully take an interest in learning from such fields how to see our own in new ways. (See Hansen 2011.)

Chapter 6

1. However, concern about diminished support for development aid is by no means a new phenomenon. Even the Pearson report from 1969 (source: http://unesdoc.unesco.org) noted that "in some donor countries, there is growing opposition or indifference to aid for development as costly, wasteful, ineffective and unnecessary."

2. Some of the details of formal organization and process mentioned here are from Justesen 2008.

3. Source: http://www.rigsrevisionen.dk/about.

4. Translated from the Danish front page of the homepage (http://www.rigsrevisionen.dk/).

5. Source: http://www.demos.co.uk.

6. Source: http://en.wikipedia.org.

7. The following is based on interviews with staff members from Department D in 2009 and 2010 and on an interview with staff members from Department A in March 2009.

8. Source: http://www.rigsrevisionen.dk.

9. Since then, the process has been made more flexible. A timeline is now decided for each individual case.

10. To be able to refer to meetings, for example, reports may make use of formulations like "The ministry has revealed. . . ." Such formulations allow auditors to express "subjective" perspectives.

11. For example, one recent case concerns a report that criticized the government for paying very high fees to private hospitals. Politicians interpreted the report as a critique of the ideological choices that led to determining the level of the fees. It was thus considered "political."

12. Source: http://www.rigsrevisionen.dk.

13. "Forster foresaw connectedness in the form of an intermediary (a participant-observer) making repeated excursions between separate realms, domains or entities, and drawing comparisons by coming to terms with the *regularities of life* practiced by each as he interprets them" (our emphasis). This is the merographic amalgam of pluralism. A postplural world has 'lost' this mathematics insofar as it can no longer enumerate such different domains." (Strathern 1992a, 215n41, quoting Nigel Rapport)

14. Meanwhile, the rigidity of procedure is never more stringent than when applied retrospectively—as suggested by the audit loop through which the veracity of this chapter was established.

Chapter 7

1. Source: http://www.thefreedictionary.com.

2. Lars' view recalls James Ferguson's description of these changes: "If the nation-building view imagined the international in the form of an idealistic UN, the state-and-society paradigm pictures a no-nonsense IMF: stern, real-world bankers, speaking what I have elsewhere called the language of economic correctness." (2006b, 82)

3. Source: http://lindaraftree.com.

References

Aanestad, Margunn. 2002. Cultivating Networks: Implementing Surgical Telemedicine. PhD dissertation, University of Oslo.

Abrahamsen, Rita. 2000. *Disciplining Democracy: Development Discourse and Good Governance in Africa*. Zed Books.

Abbate, Janet. 1999. *Inventing the Internet*. MIT Press.

Adams, Samantha, Antoinette de Bont, and Marc Berg. 2006. Looking for answers, constructing reliability: An exploration into how Dutch patients check Web-based medical information. *International Journal of Medical Informatics* 75 (1): 66–72.

Akrich, Madeleine. 1992. The de-scription of technical objects. In *Shaping Technology/Building Society: Studies in Sociotechnical Change*, ed. Wiebe Bijker and John Law. MIT Press.

Akrich, Madeleine. 1993. A gazogene in Costa Rica: An experiment in techno-sociology. In *Technological Choices: Transformation in Material Cultures since the Neolithic*, ed. Pierre Lemonnier. Taylor & Francis.

Althusser, Louis, and Étienne Balibar. 1970. *Reading Capital*. NLB.

Anders, Gerhard. 2005. Good governance as technology: Towards an ethnography of the Bretton Woods institutions. In *The Aid Effect: Giving and Governing in International Development*, ed. David Mosse and David Lewis. Pluto.

Anderson, Warwick. 2002. Introduction: Postcolonial technoscience. *Social Studies of Science* 32 (5–6): 643–658.

Aneesh, A. 2006. *Virtual Migration: The Programming of Globalization*. Duke University Press.

Asad, Talal, ed. 1973. *Anthropology and the Colonial Encounter*. Ithaca Press.

Barry, Andrew. 2001. *Political Machines: Governing a Technological Society*. Athlone.

Bauman, Zygmunt. 1990. *Thinking Sociologically*. Blackwell.

Beaulieu, Anne. 2004. Mediating ethnography: Objectivity and the making of ethnographies of the Internet. *Social Epistemology* 18 (2–3): 139–163.

Beaulieu, Anne, Sarah de Rijcke, and Bas van Heur. 2012. Authority and expertise in new sites of knowledge production. In *Virtual Knowledge*, ed. Paul Wouters, Anne Beaulieu, Andrea Scharnhorst, and Sally Wyatt. MIT Press.

Beaulieu, Anne, Andrea Scharnhorst, and Poul Wouters. 2007. Not another case study: A middle-range interrogation of ethnographic case studies in the exploration of e-science. *Science, Technology & Human Values* 32 (6): 672–693.

Beer, Stafford. 1981. *Brain of the Firm: The Managerial Cybernetics of Organization*. Wiley.

Berg, Elliott. 2000. Why aren't aid organizations better learners? In *Learning in Development Co-operation: An Introduction*, ed. Jerker Carlsson and Lennart Wohlgemuth. Almqvist & Wiksell.

Berg, Marc. 1998. The politics of technology: On bringing social theory into technological design. *Science, Technology & Human Values* 23 (4): 456–491.

Berg, Marc, and Stefan Timmermans. 2000. Order and their others: On the constitution of universalities in medical work. *Configurations* 8 (1): 31–61.

Bishop, Ann P., and Susan Leigh Star. 1996. Social informatics for digital library infrastructure and use. *Annual Review of Information Science & Technology* 31: 301–401.

Bissio, Roberto. 2001. Knowledge for development: Public good or good for the bank? *Norrag News* 29: 9–12.

Blobel, Bernd, Frank Oemig, Carolina Gonzáles, and Diego López. 2010. What is missing in health informatics standardization for phealth? *Studies in Health Technology and Informatics* 156: 3–12.

Blomberg, Jeanett, Lucy Suchman, and Randy Trigg. 1996. Reflections on a work-oriented design project. *Human-Computer Interaction* 11: 237–265.

Bloomfield, Brian P. 1995. Power, machines and social relations: Delegating to information technology in the National Health Service. *Organization* 2 (3–4): 489–518.

Bowker, Geoffrey C. 1994a. *Science on the Run: Information Management and Industrial Geophysics at Schlumberger, 1920–1940*. MIT Press.

Bowker, Geoffrey C. 1994b. Information mythology: The world of/as information. In *Information Acumen: The Understanding and Use of Knowledge in Modern Business*, ed. Lisa Bud-Frierman. Routledge.

Bowker, Geoffrey C. 1995. Second nature once removed: Time, space and representations. *Time & Society* 4 (1): 47–66.

Bowker, Geoffrey C. 2005. *Memory Practices in the Sciences*. MIT Press.

Bowker, Geoffrey C., and Susan Leigh Star. 1999. *Sorting Things Out: Classification and Its Consequences*. MIT Press.

Braudel, Ferdnand. 1976. *The Mediterranean and the Mediterranean World in the Age of Philip II*. Harper and Row.

Brown, Steven D., and Geoffrey Lightfoot. 2002. Presence, absence and accountability: E-mail and the mediation of organisational memory. In *Virtual Society? Technology, Cyberbole, Reality*, ed. Steve Woolgar. Oxford University Press.

Burrell, Jenna. 2011. User agency in the middle range: Rumors and reinvention of the Internet in Ghana. *Science, Technology & Human Values* 36 (2): 139–159.

Callon, Michel. 1998. An essay on framing and overflowing: Economic externalities revisited by sociology. In *The Laws of the Markets*, ed. Michel Callon. Blackwell.

Callon, Michel, Pierre Lascoumes, and Yannick Barthe. 2011. *Acting in an Uncertain World: An Essay on Technical Democracy*. MIT Press.

Canguilhem, Georges. 1998. The decline of the idea of progress. *Economy and Society* 27 (2/3): 313–330.

Carlsson, Jerker, and Lennart Wohlgemuth, eds. 2000. *Learning in Development Co-operation: An Introduction*. Almqvist & Wiksell.

Castells, Manuel. 1997. *The Information Age: Economy, Society and Culture*. Blackwell.

Chakravarty, Paula, and Katharine Sarikakis. 2006. *Media Policy and Globalization*. Palgrave Macmillan.

Ciborra, Claudio, ed. 2000. *From Control to Drift: The Dynamics of Corporate Information Infrastructures*. Oxford University Press.

Ciborra, Claudio, and Angelo Failla. 2000. Infrastructure as a process: The case of CRM in IBM. In *From Control to Drift: The Dynamics of Corporate Information Infrastructures*, ed. Claudio Ciborra. Oxford University Press.

Ciborra, Claudio, and Ole Hanseth. 2000. Introduction. In *From Control to Drift: The Dynamics of Corporate Information Infrastructures*, ed. Claudio Ciborra. Oxford University Press.

Clarke, Adele E., and Joan H. Fujimura, eds. 1992. *The Right Tools for the Job: At Work in Twentieth-Century Life Sciences*. Princeton University Press.

Cooke, Bill, and Uma Kothari. 2001. *Participation: The New Tyranny?* Zed Books.

Cooper, Frederick, and Randall M. Packard. 1997a. Introduction. In *International Development and the Social Sciences: Essays on the History and Politics of Knowledge*, ed. Frederick Cooper and Randall M. Packard. University of California Press.

Cooper, Frederick, and Randall M. Packard, eds. 1997b. *International Development and the Social Sciences: Essays on the History and Politics of Knowledge*. University of California Press.

Corsin-Jiménez, Alberto. 2007. *The Anthropology of Organisations. Aldershot.* Ashgate.

Cowen, Michael P., and Robert W. Shenton. 1996. *Doctrines of Development*. Routledge.

Crapanzano, Vincent. 1992. *Hermes' Dilemma and Hamlet's Desire: On the Epistemology of Interpretation*. Harvard University Press.

Curtis, Brian. 2002. *The Politics of Population: State Formation, Statistics, and the Census of Canada, 1840–1875*. University of Toronto Press.

Dalakoglou, Dimitris. 2010. The road: An ethnography of the Albanian-Greek cross-border motorway. *American Ethnologist* 37 (1): 132–149.

Daston, Lorraine, ed. 2000. *Biographies of Scientific Objects*. University of Chicago Press.

Dean, Mitchell. 1999. *Governmentality: Power and Rule in Modern Society*. Sage.

Deleuze, Gilles. 1995. *Negotiations 1972–1990*. Columbia University Press.

Deleuze, Gilles. 2007. *Two Regimes of Madness: Texts and Interviews 1975–1995*. Semiotext(e).

Deleuze, Gilles, and Félix Guattari. 1987. *A Thousand Plateaus: Capitalism and Schizophrenia*. University of Minnesota Press.

de Sardan, Olivier. Jean-Pierre. 2005. Anthropology and Development: Understanding Contemporary Social Change. Zed Books.

Desrosiéres, Alain. 1998. *The Politics of Large Numbers*. Harvard University Press.

Dormans, Stefan, and Jan Kok. 2010. An alternative approach to large historical databases. Exploring best practices with collaboratories. *Historical Methods* 43 (3): 97–107.

Drayton, Richard. 2000. *Nature's Government: Science, Imperial Britain, and the "Improvement" of the World*. Yale University Press.

Duerr, Hans Peter. 1985. *Dreamtime: Concerning the Boundary between Wilderness and Civilization*. Blackwell.

Du Gay, Paul, and Graham Salaman. 1992. The cult(ure) of the customer. *Journal of Management Studies* 29 (5): 615–633.

Easterly, William. 2002. What Did Structural Adjustment Adjust? The Association of Policies and Growth with Repeated IMF and World Bank Adjustment Loans. Working Paper 11, Center for Global Development. Available at http://www.cgdev.org.

Easterly, William. 2006. *The White Man's Burden: Why the West's Efforts to Aid the Rest Have Done So Much Ill and So Little Good.* Oxford University Press.

Edelman, Marc, and Angelique Haugerud. 2005. Introduction: The anthropology of development and globalization. In *The Anthropology of Development and Globalization: From Classical Political Economy to Contemporary Neoliberalism*, ed. Marc Edelman and Angelique Haugerud. Blackwell.

Edgren, Gus. 2000. Fashions, myths and delusion: Obstacles to organizational learning in aid agencies. In *Learning in Development Co-operation: An Introduction*, ed. Jerker Carlsson and Lennart Wohlgemuth. Almqvist & Wiksell.

Edwards, Mike. 1994. NGOs in the age of information. *IDS Bulletin* 25 (2): 117–125.

Edwards, Paul N. 2003. Infrastructure and modernity: Force, time and social organization in the history of sociotechnical systems. In *Modernity and Technology*, ed. Thomas J. Misa, Paul Brey, and Andrew Feenberg. MIT Press.

Edwards, Paul N. 2010. *A Vast Machine: Computer Models, Climate Data, and the Politics of Global Warming.* MIT Press.

Edwards, Paul N., Steven J. Jackson, Geoffrey C. Bowker, and Cory P. Knobel. 2007. Understanding infrastructure: Dynamics, tensions, and design. *First Monday* 12 (6). Available at http://firstmonday.org.

Edwards, Paul N., Geoffrey C. Bowker, Steven J. Jackson, and Robin Williams. 2009. Introduction: An agenda for infrastructure studies. *Journal of the Association for Information Systems* 10 (5): 364–374.

Egyedi, Tineke M., and Donna Mehos, eds. 2012. *Inverse Infrastructures: Disrupting Networks from Below.* Edward Elgar.

Elyachar, Julia. 2010. Phatic labor, infrastructure, and the question of empowerment in Cairo. *American Ethnologist* 37 (3): 452–464.

Escobar, Arturo. 1995. *Encountering Development: The Making and Unmaking of the Third World.* Princeton University Press.

Evans-Pritchard, E. E. 1951. *Social Anthropology.* Cohen and West.

Ferguson, James. 1990. *The Anti-Politics Machine: "Development," Depoliticization, and Bureaucratic Power in Lesotho.* Cambridge University Press.

Ferguson, James. 1997. Anthropology and its evil twin: "Development" in the constitution of a discipline. In *International Development and the Social Sciences: Essays on the History and Politics of Knowledge*, ed. Frederick Cooper and Randall M. Packard. University of California Press.

Ferguson, James. 1999. *Expectations of Modernity: Myths and Meanings of Urban Life on the Zambian Copperbelt.* University of California Press.

Ferguson, James. 2006a. *Global Shadows: Africa in the Neoliberal World Order*. Duke University Press.

Ferguson, James. 2006b. Transnational topographies of power: Beyond "the State" and "Civil Society" in the study of African politics. In *Accelerating Possession: Global Futures of Property and Personhood*, ed. Bill Maurer and Gabriele Schwab. Columbia University Press.

Finnemore, Martha. 1997. Redefining development at the World Bank. In *International Development and the Social Sciences: Essays on the History and Politics of Knowledge*, ed. Frederick Cooper and Randall M. Packard. University of California Press.

Flichy, Patrice. 2007. *The Internet Imaginaire*. MIT Press.

Forsythe, Diana. 2001. *Studying Those Who Study Us: An Anthropologist in the World of Artificial Intelligence*. Stanford University Press.

Frank, Leonard. 1997. The development game. In *The Post-Development Reader*, ed. Majid Rahnema and Victoria Bawtree. Zed Books.

Franzen, Jonathan. 2002. *How to Be Alone*. Fourth Estate.

Gad, Christopher, and Casper Bruun Jensen. 2010. On the consequences of post-ANT. *Science, Technology & Human Values* 35 (1): 55–80.

Gad, Christopher, and David Ribes. Forthcoming. Editorial introduction. *Science, Technology & Human Values*.

Gardner, Katy, and David Lewis. 1996. *Anthropology, Development and the Postmodern Challenge*. Pluto.

Garfinkel, Harold. 1967. *Studies in Ethnomethodology*. Prentice-Hall.

Gefou-Medianou, Dimitra. 2000. Disciples, discipline and reflection: Anthropological encounters and trajectories. In *Audit Cultures: Anthropological Studies in Accountability, Ethics and the Academy*, ed. Marilyn Strathern. Routledge.

Goldman, Michael. 2005. *Imperial Nature: The World Bank and Struggles for Social Justice in the Age of Globalization*. Yale University Press.

Gould, Jeremy. 2005. Timing, scale and style: Capacity as governmentality in Tanzania. In *The Aid Effect: Giving and Governing in International Development*, ed. David Mosse and David Lewis. Pluto.

Gusterson, Hugh. 1997. Studying-up revisited. *Political and Legal Anthropology Review* 20 (1): 114–119.

Hacking, Ian. 1986. Making up people. In *Reconstructing Individualism: Autonomy, Individuality and the Self in Western Thought*, ed. Thomas C. Heller, David C. Wellbery, and Morton Sosna. Stanford University Press.

Halewood, Michael. 2005. A. N. Whitehead, information and social theory. *Theory, Culture & Society* 22 (6): 73–94.

Hansen, Birgitte Gorm. 2011. Adapting in the Knowledge Economy: Lateral Strategies for Scientists and Those Who Study Them. PhD dissertation, Copenhagen Business School.

Hanseth, Ole. n.d. From Systems and Tools to Networks and Infrastructures—from Design to Cultivation: Towards a Theory of ICT Solutions and Its Design Methodology Implications. Available at http://heim.ifi.uio.no.

Haraway, Donna. 1997. *Modest_Witness@Second_Millennium.FemaleMan©_Meets_OncoMouse™ Feminism and Technoscience*. Routledge.

Harper, Richard. 1998. *Inside the IMF: An Ethnography of Documents, Technology, and Organisational Action*. Academic Press.

Harvey, Penny. 2010. Cementing relations: The materiality of roads and public spaces in provincial Peru. *Social Analysis* 54 (2): 28–46.

Heath, Deborah. 1998a. Locating genetic knowledge: Picturing Marfan Syndrome and its traveling constituencies. *Science, Technology & Human Values* 23 (1): 71–97.

Heath, Deborah. 1998b. Bodies, antibodies and modest interventions. In *Cyborgs and Citadels: Anthropological Interventions in Emerging Sciences and Technologies*, ed. Gary L. Downey, Joseph Dumit, and Sharon Traweek. School of American Research Press.

Hemment, Julie. 2004. The riddle of the third sector: Civil society, international aid, and NGOs in Russia. *Anthropological Quarterly* 77 (2): 215–241.

Hine, Christine. 2000. *Virtual Ethnography*. Sage.

Hine, Christine, ed. 2005. *Virtual Methods: Issues in Social Research on the Internet*. Berg.

Ho, Karen. 2009. *Liquidated: An Ethnography of Wall Street*. Duke University Press.

Hutchins, Edwin. 1995. *Cognition in the Wild*. MIT Press.

Jasanoff, Sheila, ed. 2004. *States of Knowledge: The Co-Production of Science and Social Order*. Routledge.

Jensen, Casper Bruun. 2004. A non-humanist disposition. On performativity, practical ontology and intervention. *Configurations* 12 (2): 229–261.

Jensen, Casper Bruun. 2007a. Infrastructural fractals: Revisiting the micro-macro distinction in social theory. *Environment and Planning. D, Society & Space* 25 (5): 832–850.

Jensen, Casper Bruun. 2007b. Sorting attachments: Usefulness of STS in health care policy and practice. *Science as Culture* 16 (3): 237–251.

Jensen, Casper Bruun. 2008. Power, technology and social studies of health care: An infrastructural inversion. *Health Care Analysis* 16 (4): 355–374.

Jensen, Casper Bruun. 2010a. *Ontologies for Developing Things: Making Health Care Futures Through Technology*. Sense.

Jensen, Casper Bruun. 2010b. Asymmetries of knowledge: Virtual ethnography and ICT for development. *Methodological Innovations Online* 5 (1): 72–85.

Jensen, Casper Bruun. 2012. Motion: The task of anthropology is to invent relations. *Critique of Anthropology* 32 (1): 47–53.

Jensen, Casper Bruun. Forthcoming. Continuous variations: The conceptual and the empirical in STS. *Science, Technology and Human Values* (special issue on The Conceptual and the Empirical).

Jensen, Casper Bruun, and Christopher Gad. 2009. Philosophy of technology as empirical philosophy: Comparing technological scales in practice. In *New Waves in Philosophy of Technology*, ed. Jan-Kyrre Berg Olsen, Søren Riis, and Evan M. Selinger. Palgrave MacMillan.

Jensen, Casper Bruun, and Kjetil Rödje, eds. 2010. *Deleuzian Intersections: Science, Technology, Anthropology*. Berghahn.

Justesen, Lise. 2008. Kunsten at skrive revisionsrapporter. En beretning om forvaltningsrevisionens beretninger. PhD dissertation, Copenhagen Business School.

Keck, Margaret A., and Kathryn Sikkink. 1998. *Activists Beyond Borders: Advocacy Networks in International Politics*. Cornell University Press.

Knorr Cetina, Karin and Urs Bruegger. 2002. Global microstructures: The virtual societies of financial markets. *American Journal of Sociology* 107 (4): 905–950.

Kockelman, Paul. 2010. Enemies, parasites, and noise: How to take up residence in a system without becoming a term in it. *Journal of Linguistic Anthropology* 20 (2): 406–421.

Langwick, Stacey A. 2007. Devils, parasites and fierce needles: Healing and the politics of translation in Southern Tanzania. *Science, Technology & Human Values* 32 (1): 88–117.

Larner, Wendy, and William Walters, eds. 2004. *Global Governmentality: Governing International Spaces*. Routledge.

Latour, Bruno. 1987. *Science in Action: How to Follow Scientists and Engineers Through Society*. Harvard University Press.

Latour, Bruno. 1988. *The Pasteurization of France (with "Irreductions")*. Cambridge University Press.

Latour, Bruno. 1993. *We Have Never Been Modern*. Harvester Wheatsheaf.

Latour, Bruno. 1996. *Aramis, or the Love of Technology*. Harvard University Press.

Latour, Bruno. 1999. *Pandora's Hope: Essays on the Reality of Science Studies*. Harvard University Press.

Latour, Bruno. 2004. *Politics of Nature: How to Bring the Sciences Into Democracy*. Harvard University Press.

Latour, Bruno, and Peter Weibel, eds. 2005. *Making Things Public: Atmospheres of Democracy*. MIT Press.

Lewis, David. 1998. Partnership as process: Building an institutional ethnography of an inter-agency aquaculture project in Bangladesh. In *Development as Process: Concepts and Methods for Working with Complexity*, ed. David Mosse, John Farrington, and Alan Rew. Routledge.

Lewis, David, and Nazneen Kanji. 2009. *Non-Governmental Organizations and Development*. Routledge.

Lewis, David, and David Mosse. 2006. *Development Brokers and Translators: The Ethnography of Aid and Agencies*. Kumarian.

Li, Tania M. 1999. *Transforming the Indonesian Uplands: Marginality, Power and Production*. Harwood.

Marcus, George. 1995. Ethnography in/of the world system: The emergence of multi-sited ethnography. *Annual Review of Anthropology* 24: 95–117.

Maurer, Bill. 2005. *Mutual Life, Limited: Islamic Banking, Alternative Currencies, Lateral Reason*. Princeton University Press.

Maurer, Bill. 2006. *Pious Property: Islamic Mortgages in the United States*. Russell Sage Foundation.

McLuhan, Marshall. 1997. *Understanding Media: The Extensions of Man*. MIT Press.

Mol, Annemarie. 2002. *The Body Multiple: Ontology in Medical Practice*. Duke University Press.

Moon, Samuel, and Tim Williamson. 2010. Greater aid transparency: Crucial for aid effectiveness. *ODI Project Briefing* 35: 1–4.

Moore, Sally Falk. 1987. Explaining the present: Theoretical dilemmas in processual ethnography. *American Ethnologist* 14 (4): 727–736.

Moser, Ingunn, and John Law. 2006. Fluids or flows? Information and qualculation in medical practice. *Information Technology & People* 19 (1): 55–73.

Mosse, David. 2004. Is good policy unimplementable? Reflections on the ethnography of aid policy and practice. *Development and Change* 35 (4): 639–671.

Mosse, David. 2005a. Global governance and the ethnography of international aid. In *The Aid Effect: Giving and Governing in International Development*, ed. David Mosse and David Lewis. Pluto.

Mosse, David. 2005b. *Cultivating Development: An Ethnography of Aid Policy and Practice*. Pluto.

Mosse, David, and David Lewis, eds. 2005. *The Aid Effect: Giving and Governing in International Development*. Pluto.

Mosse, David, and David Lewis. 2006. Theoretical approaches to brokerage and translation in development. In *Brokers and Translators: The Ethnography of Aid and Agencies*, ed. David Lewis and David Mosse. Kumarian.

Mouffe, Chantal. 2005. *On the Political*. Routledge.

Munro, Rolland, and Jan Mouritsen, eds. 1996. *Accountability: Power, Ethos and the Technologies of Managing*. International Thomson Business Press.

Nader, Laura, ed. 1996. *Naked Science: Anthropological Inquiry into Boundaries, Power and Knowledge*. Routledge.

Nielsen, Morten. 2012. Roadside inventions: Making time and money work at a road construction site in Mozambique. *Mobilities* 7 (4): 467–480.

Ong, Aihwa, and Stephen J. Collier, eds. 2005. *Global Assemblages: Technology, Politics, and Ethics as Anthropological Problems*. Blackwell.

Orr, Jackie. 2000. The ecstasy of miscommunication: Cyberpsychiatry and mental dis-ease. In *Doing Science + Culture: How Cultural and Interdisciplinary Studies are Changing the Way We Look at Science and Medicine*, ed. Roddey Reid and Sharon Traweek. Routledge.

Oudshoorn, Nelly, and Trevor Pinch, eds. 2003. *How Users Matter: The Co-Construction of Users and Technology*. MIT Press.

Oudshoorn, Nelly, Els Rommes, and Marcelle Stienstra. 2004. Configuring the user as everybody: Gender and design cultures in information and communication technologies. *Science, Technology & Human Values* 29 (1): 30–63.

Pearson, Lester B. 1969. *Partners in Development: Report of the Commission on International Development*. Praeger.

Pels, Peter. 2000. The trickster's dilemma: Ethics and the technologies of the anthropological self. In *Audit Cultures: Anthropological Studies in Accountability, Ethics and the Academy*, ed. Marilyn Strathern. Routledge.

Peluso, Nancy L. 1991. *Rich Forests, Poor People: Resource Control and Resistance in Java*. University of California Press.

Pickering, Andrew. 1995. *The Mangle of Practice: Time, Agency and Science*. University of Chicago Press.

Pirnejad, Habibollah, Roland Bal, and Marc Berg. 2008. Building an inter-organizational communication network and challenges for preserving interoperability. *International Journal of Medical Informatics* 77 (12): 818–827.

Pollock, Neil, and Robin Williams. 2010. E-infrastructures: How do we know and understand them? Strategic ethnography and the biography of artefacts. *Computer Supported Cooperative Work* 19 (6): 521–556.

Porter, Theodore M. 1995. *Trust in Numbers: The Pursuit of Objectivity in Science and Public Life*. Princeton University Press.

Power, Marcus. 2003. *Rethinking Development Geographies*. Routledge.

Power, Michael. 1994. *The Audit Explosion*. Demos. Available at http://www.demos.co.uk.

Power, Michael. 1997. *The Audit Society: Rituals of Verification*. Oxford University Press.

Power, Michael. 2000. *The Audit Implosion: Regulating Risk From the Inside*. ICAEW.

Power, Michael. 2007. *Organizing Uncertainty: Designing a World of Risk Management*. Oxford University Press.

Ribes, David, and Thomas A. Finholt. 2009. The long now of technology infrastructure: Articulating tensions in development. *Journal of the Association for Information Systems* 10 (5). Available at http://aisel.aisnet.org.

Richards, Thomas. 1993. *The Imperial Archive: Knowledge and the Fantasy of Empire*. Verso.

Riles, Annelise. 2000. *The Network Inside Out*. University of Michigan Press.

Riles, Annelise. 2005. The network inside out. In *The Anthropology of Development and Globalization: From Classical Political Economy to Contemporary Neoliberalism*, ed. Marc Edelman and Angelique Haugerud. Blackwell.

Riles, Annelise. 2006a. [Deadlines]: Removing the brackets on politics in bureaucratic and anthropological analysis. In *Documents: Artefacts of Modern Knowledge*, ed. Annelise Riles. University of Michigan Press.

Riles, Annelise. 2006b. Introduction: In response. In *Documents: Artefacts of Modern Knowledge*, ed. Annelise Riles. University of Michigan Press.

Riles, Annelise, ed. 2006c. *Documents: Artefacts of Modern Knowledge*. University of Michigan Press.

Riles, Annelise. 2011. *Collateral Knowledge: Legal Reasoning in the Global Financial Markets*. University of Chicago Press.

Robins, Kevin, and Frank Webster. 1999. *Times of the Technoculture: From the Information Society to Virtual Life*. Routledge.

Rod, Morten Hulvej. 2010. Forebyggelsens momenter: En antropologisk analyse af evidens og etik i tiltag rettet mod unges brug af alkohol. PhD dissertation, University of Copenhagen.

Rogers, Everett, ed. 2003. *Diffusion of Innovations*. Free Press.

Rose, Nikolas. 1999. *Powers of Freedom: Reframing Political Thought*. Cambridge University Press.

Rostow, Walt W. 1960. *The Stages of Economic Growth: A Non-Communist Manifesto*. Cambridge University Press.

Rottenburg, Richard. 2009. *Far-Fetched Facts: A Parable of Development aid*. MIT Press.

Sahay, Sundeep, Eric Monteiro, and Margunn Aanestad. 2009. Configurable politics and asymmetric integration: Health e-infrastructures in India. *Journal of the Association for Information Systems* 10 (5): 399–414.

Schmidt, Susanne K., and Raymund Werle. 1998. *Coordinating Technology: Studies in the International Standardization of Telecommunications*. MIT Press.

Schwartzman, Helen B. 1989. *The Meeting: Gatherings in Organizations and Communities*. Plenum.

Scott, James C. 1998. *Seeing Like a State: How Certain Schemes to Improve the Human Condition Have Failed*. Yale University Press.

Shore, Cris, and Susan Wright, eds. 1997. *Anthropology of Policy: Critical Perspectives on Governance and Power*. Routledge.

Simon, Herbert A. 1957. *Models of Man: Social and Rational*. Wiley.

Smith, Barbara Herrnstein. 1997. *Belief and Resistance: Dynamics of Contemporary Intellectual Controversy*. Duke University Press.

Star, Susan Leigh, ed. 1995a. *The Cultures of Computing*. Blackwell Publishers.

Star, Susan Leigh. 1995b. The politics of formal representation: Wizards, gurus and organizational complexity. In *Ecologies of Knowledge: Work and Politics in Science and Technology*, ed. Susan Leigh Star. State University of New York Press.

Star, Susan Leigh, and Karen Ruhleder. 1996. Steps toward an ecology of infrastructure: Design and access for large information spaces. *Information Systems Research* 7 (1): 111–134.

Star, Susan Leigh, and Anselm Strauss. 1999. Layers of silence, arenas of voice: The ecology of visible and invisible work. *Computer Supported Cooperative Work* 8: 9–30.

Stengers, Isabelle. 2011a. Comparison as a matter of concern. *Common Knowledge* 17 (1): 48–64.

Stengers, Isabelle. 2011b. *Thinking With Whitehead: A Free and Wild Creation of Concepts*. Harvard University Press.

Stiglitz, Joseph E. 2002. *Globalization and Its Discontents*. Norton.

Strathern, Marilyn. 1988. *The Gender of the Gift: Problems with Women and Problems with Society in Melanesia*. University of California Press.

Strathern, Marilyn. 1989. Between a Melanesianist and a deconstructive feminist. *Australian Feminist Studies* 10: 49–69.

Strathern, Marilyn. 1991. *Partial Connections*. Rowman and Littlefield.

Strathern, Marilyn. 1992a. *After Nature: English Kinship in the Late Twentieth Century*. Cambridge University Press.

Strathern, Marilyn. 1992b. Parts and wholes: Refiguring relationships in a post-plural world. In *Conceptualizing Society*, ed. Adam Kuper. Routledge.

Strathern, Marilyn. 1999. *Property, Substance and Effect. Anthropological Essays on Persons and Things*. Athlone.

Strathern, Marilyn, ed. 2000a. *Audit Cultures: Anthropological Studies in Accountability, Ethics and the Academy*. Routledge.

Strathern, Marilyn. 2000b. Introduction: New accountabilities. In *Audit Cultures: Anthropological Studies in Accountability, Ethics and the Academy*, ed. Marilyn Strathern. Routledge.

Strathern, Marilyn. 2000c. Afterword: Accountability . . . and ethnography. In *Audit Cultures: Anthropological Studies in Accountability, Ethics and the Academy*, ed. Marilyn Strathern. Routledge.

Strathern, Marilyn. 2000d. The tyranny of transparency. *British Educational Research Journal* 26 (3): 309–321.

Strathern, Marilyn. 2006. Bullet-proofing: A tale from the United Kingdom. In *Documents: Artefacts of Modern Knowledge*, ed. Annelise Riles. University of Michigan Press.

Suchman, Lucy. 1987. *Plans and Situated Actions: The Problem of Human-Machine Communication*. Cambridge University Press.

Suchman, Lucy. 1999. "I have, more than ever, a sense of the immovability of these institutions": Conversation with Lucy Suchman. Available at http://www.iwp.jku.at.

Suchman, Lucy. 2006. *Human-Machine Reconfigurations: Plans and Situated Actions*. Cambridge University Press.

Suchman, L. 2011. Anthropological relocations and the limits of design. *Annual Review of Anthropology* 40: 1–18.

Terranova, Tiziana. 2004. *Network Culture: Politics for the Information Age*. Pluto.

Thévenot, Laurent. 1984. Rules and implements: Investment in forms. *Social Sciences Information. Information Sur les Sciences Sociales* 23 (1): 1–45.

Toffler, Alvin. 1970. *Future Shock*. Random House.

Traweek, Sharon. 1988. *Beamtimes and Lifetimes: The World of High-Energy Physicists*. Harvard University Press.

Tsing, Anna. 2005. *Friction: An Ethnography of Global Connection*. Princeton University Press.

Vaughan, Diana. 1996. *The Challenger Launch Decision: Risky Technology, Culture, and Deviance at NASA*. University of Chicago Press.

Verran, Helen. 2001. *Science and an African Logic*. University of Chicago Press.

Verran, Helen. 2002. A postcolonial moment in science studies: Alternative firing regimes of environmental scientists and aboriginal landowners. *Social Studies of Science* 32: 729–762.

Verran, Helen. 2011. Imagining nature politics in the era of Australia's emerging markets in environmental services interventions. *Sociological Review* 59 (3): 411–431.

Vikkelsø, Signe. 2005. Subtle redistribution of work, attention and risks: Electronic patient records and organizational consequences. *Scandinavian Journal of Informatics* 17 (1): 3–30.

Viveiros de Castro, Eduardo. 2004. Exchanging perspectives: The transformation of objects into subjects in Amerindian ontologies. *Common Knowledge* 10 (3): 463–484.

Wagner, Roy. 1991. The fractal person. In *Big Men and Great Men: Personifications of Power in Melanesia*, ed. Maurice Godelier and Marilyn Strathern. Cambridge University Press.

Wapenhans, Willi A. 2000. Learning by doing: Reflections on 35 years with the World Bank. In *Learning in Development Co-operation: An Introduction*, ed. Jerker Carlsson and Lennart Wohlgemuth. Almqvist & Wiksell.

Wieslander, Anna. 2000. When do we ever learn? In *Learning in Development Co-operation: An Introduction*, ed. Jerker Carlsson and Lennart Wohlgemuth, 250–268. Almqvist & Wiksell.

Williams, Raymond. 1981. *Keywords: A Vocabulary of Culture and Society*. Fontana Books.

Winters, Matthew. 2009. Accountability, Participation and Foreign Aid Effectiveness. Niehaus Center for Globalization and Governance, Princeton University. Available at http://www.gsdrc.org.

Winthereik, Brit Ross. 2003. "We fill in our working understanding": On codes, classifications and the production of accurate data. *Methods of Information in Medicine* 42: 489–496.

Winthereik, Brit Ross. 2010. The project multiple: Enactments of systems development. *Scandinavian Journal of Information Systems* 22 (2): 49–64.

Winthereik, Brit Ross, Nis Johannsen, and Dixi Louise Strand. 2008. Making technology public: Challenging the notion of script through an e-health demonstration video. *Information Technology & People* 21: 116–132.

Winthereik, Brit Ross, and Henriette Langstrup. 2010. When patients care (too much) for information. In *Care Practices: On Tinkering in Clinics, Homes and Farms*, ed. Annemarie Mol, Ingunn Moser, and Jeanett Pols. Transcript Verlag.

Winthereik, Brit Ross, and Helen Verran. 2012. Ethnographic stories as generalizations that intervene. *Science Studies* 25 (1): 37–51.

Winthereik, Brit Ross, and Signe Vikkelsø. 2005. ICT and integrated care: Some dilemmas of standardising inter-organisational communication. *Computer Supported Cooperative Work* 14 (1): 43–67.

Wong, Kwok-fu. 2003. Empowerment as a panacea for poverty: Old wine in new bottles? Reflections on the World Bank's conception of power. *Progress in Development Studies* 3 (4): 307–322.

Woolgar, Steve. 1991. Configuring the user: The case of usability trials. In *A Sociology of Monsters: Essays on Power, Technology and Domination*, ed. John Law. Routledge.

Wouters, Paul, Katie Vann, Andrea Scharnhorst, Matt Ratto, Iina Hellsten, Jenny Fry, and Anne Beaulieu. 2008. Messy shapes of knowledge—STS explores informatization, new media, and academic work. In *Handbook of Science and Technology studies*, ed. Edward Hackett, Olga Amsterdamska, Michael Lynch, and Judy Wajcman. MIT Press.

Wyatt, Sally, and Brian Balmer. 2007. Home on the range: What and where is the middle in science and technology studies? *Science, Technology & Human Values* 32 (6): 619–626.

Yarrow, Thomas, and Soumhya Venkatesan, eds. 2012. *Differentiating Development: Beyond an Anthropology of Critique*. Berghahn.

Ya'u, Yunusa Z. 2005. Globalisation, ICTs, and the new imperialism: Perspectives on Africa in the global electronic village. *Africa Development* 20 (1/2): 98–124.

Zhan, Mei. 2009. *Other-Worldly: Making Chinese Medicine Through Transnational Frames*. Duke University Press.

Zuboff, Shoshana. 1989. *In the Age of the Smart Machine*. Basic Books.

Index